The Tricentennial People

HUMAN APPLICATIONS OF THE NEW GENETICS

The Tricentennial People

HUMAN APPLICATIONS OF THE NEW GENETICS

The Tricentennial People

HUMAN APPLICATIONS OF THE NEW GENETICS

Iowa State University Press / Ames, Iowa

1 9 7 8

THIS VOLUME is an outgrowth of a symposium sponsored by Clarke College, Dubuque, Iowa, October 21, 1975. The symposium was centered on the biological, legal, moral, and psychological effects of the use of the new genetic information. It was one in a series of Clarke-sponsored symposia on contemporary topics including "Liturgy and Architecture" in 1964, "Creative America" in 1965, "Man in a Man-Made World" (effect of technology and behavioral mechanisms on the human person) in 1968, and "Creative Dissent" (political currents of the 1960s) in 1971. Clarke offers these symposia as a service to the college communities of Dubuque and to the city. They bring together scholars who engage in open interchange on topics of immediate concern and awaken students to the complexities of the larger world.

Volume Editor: **Sister Marguerite Neumann,** Professor of Chemistry, Clarke College, and Chairman of the Symposium Committee

© 1978 The Iowa State University Press
All rights reserved

Composed and printed by
The Iowa State University Press
Ames, Iowa 50010

No part of this publication may be reproduced, stored in a retrieval system, or transmitted, in any form or by any means—electronic, mechanical, photocopying, recording, or otherwise—without the prior written permission of the publisher.

First edition, 1978

Library of Congress Cataloging in Publication Data
Main entry under title:

The Tricentennial people.

"An outgrowth of a symposium sponsored by Clarke College, Dubuque, Iowa, October 21, 1975."
  Includes bibliographical references and index.
  1. Human genetics—Social aspects—Congresses.
2. Genetic engineering—Social aspects—Congresses.
I. Neumann, Marguerite, 1914- II. Clarke College, Dubuque, Iowa.
[DNLM: 1. Genetics, Human—Congresses. QH431 T823 1975]
QH431.T7   301.24'3   78-8420
ISBN 0-8138-1650-5

Contents

	Foreword	vii
	Prologue	ix
1/	Genetics and the Biological Basis of the Human Condition *Elof Axel Carlson, Ph.D.*	3
2/	Genetic Counseling: Boon or Bane? *Robert F. Murray, Jr., M.D.*	29
3/	Genetics and the Law *Margery W. Shaw, M.D., J.D.*	48
4/	Genetics, Reproductive Biology, and Bioethics *LeRoy Walters, Ph.D.*	66
5/	General Comment	81
	Epilogue	97
	Index	101

Foreword

THE TOPIC of human genetics is receiving more attention than ever before in professional literature and universities, government, and other groups charged with the responsibility of disseminating knowledge for the common good. Certainly contemporary genetics is a rapidly growing part of science—one filled with exciting discoveries. It is more than a scientific frontier. Its discoveries suggest new applications to human life. Can we learn more about the health of a child in the womb? Can we plan the genetic makeup of individuals or of the whole tricentennial people? Such new applications, furthermore, suggest new choices. If we can plan, if we can control, should we? We believe values are clarified when they are tested against new challenges. Open discussion with its accompanying diversity of opinions can be an exciting way to clarify personal values.

Robert Giroux
President
Clarke College

Prologue

PERHAPS more than ever before, people of this generation are looking forward to the next hundred years. While it seemed to have been the custom in the past to look to a "Golden Age" that existed at some time before their own existence, e.g., The Garden of Eden, today many are looking to the future for a "Golden Age." It would be an age not given to them from on high by a superior power but rather one they would have shaped with their own hands.

Notwithstanding the more optimistic among humans, there are some who are not so sure that the human race has the wisdom to achieve such a state of well-being. What are the possibilities? Futurology as a scientific discipline has not yet developed to the point it can say with much assurance that a "Golden Age" is inevitable if we were to act wisely now, or that it would not be attained were we to act foolishly. There are too many contingencies. Yet there is one fact that is beyond question: Man has traveled a long road in a few hundred thousand years, even at a time when the degree of control over his environment and his own body was very limited. Now that humans have the knowledge and power to control many of the processes heretofore left to nature, will they be able to use that power radically to improve their environment and themselves? Recent awareness of ecological irresponsibility has sobered men and women everywhere. At one time, H. J. Muller was optimistic about the power of man to improve himself through the application of eugenics, as Elof Carlson has remarked [Elof Carlson, "Eugenics Revisited: The Case for Germinal Choice," *Stadler Symposium* 5:13-34, 1973], but soon found that people and governments were not ready.

Still, if we do nothing, the prospects as seen by some are not enticing. Dr. Bentley Glass observed in a paper, "Human Heredity and Ethical Problems," first presented in 1970 but later published in 1972, that

> . . . to contemplate the man of tomorrow who must begin his day by adjusting his spectacles and his hearing aid, inserting his

false teeth, taking an allergy injection in one arm and an insulin injection in the other, and topping off his preparations for life by taking a tranquilizing pill, is none too pleasant. [*Perspectives in Biology and Medicine* 15(2), winter, 1972. Reprinted by Society for Health and Human Values, 1975, p. 10, University of Chicago.]

Of course, this is assuming no major catastrophe such as a nuclear war.

Society today seems to be in the situation where it possesses powerful tools but does not yet know what it wishes to do with them. There are no clear social goals agreed upon. Nor is there a consensus regarding the values we wish to promote; that will probably be our most difficult task. Particularly important, as Etzioni has pointed out, is that no one category of citizens or professions can make the determination; as broad a base of discussion as possible is required. The whole body of citizens should be given the opportunity to make an input into the decision-making process. (Amitai Etzioni, *Genetic Fix*, New York: Macmillan Publishing Company, 1973.)

To focus our present discussion we need to limit it to human beings: What kind of a human being would we want to fashion? One course of action would be to identify the respective heroes of various pieces of literature (taken from different cultures and epochs) and select the elements common to all. In a sense, this would describe the universal hero, the noble person whom most humans admire. From religious convictions some persons would suggest as the perfect type of humanity particular individuals like Jesus, Moses, or Buddha. In a more specific vein, I want to cite two suggestions made by eminent scientists (*Perspectives in Biology and Medicine* reprint, p. 15):

> I once suggested [Bentley Glass, *Science and Liberal Education*, Baton Rouge: Louisiana State University Press, 1959] that we might agree upon such goals as "freedom from gross physical or mental defects, sound health, high intelligence, general adaptability, integrity of character, and mobility of spirit." I did not imply that these characteristics were in any case fully or even partially genetic in nature.

> H. J. Muller [H. J. Muller, in H. Hoagland and R. W. Burhoe (eds.), *Evolution and Man's Progress*, New York: Columbia University Press, 1962] selected a different list: "genuine warmth of fellow feeling and a cooperative disposition, a depth and breadth of intellectual capacity, moral courage and integrity, and appreciation of nature and art, and

PROLOGUE

an aptness of expression and of communication"; on the physical side, "to better the genetic foundations of health, vigor, and longevity; to reduce the need for sleep; to bring the induction of sedation and stimulation under more voluntary control; and to develop increasing physical tolerances and aptitudes in general."

But having said all this, who decides what kind of human being we would want to clone or "manufacture" and by what process can such be decided?

An objective of this symposium on "The Tricentennial People" is to reflect on some of these issues—scientific, ethical, legal, and social. We need to know

> What the exact *relationships* are between genes and disease/defects; between genes and what we consider to be the positive and desirable attributes; between genes and the environment—internal and external.
>
> How to correct or prevent undesirable phenotypic structure and behavior (assuming we have defined what is undesirable) and how to promote the desired positive attributes.
>
> How, through wide public discussion, to achieve socially agreed-upon genetic goals.
>
> How to identify accurately and acceptably those individuals who are genetically "at risk."
>
> How to counsel such persons in a manner respectful of their human dignity.
>
> How to identify the precise ethical issues and how to resolve conflict values constructively.
>
> How to formulate laws that will best aid in attaining socially agreed-upon goals.
>
> How to deal justly with ethical dissidents to specific genetic laws.

These are all-embracing questions to which adequate answers will be forthcoming only in the future. However, the present contributors, each a leader in his or her respective area, will provide data, insights, and attitudes as bases for further research and reflection.

Albert Moraczewski
Chairman, Panel

The Tricentennial People

HUMAN APPLICATIONS OF THE NEW GENETICS

1/
Genetics and the Biological Basis of the Human Condition

ELOF AXEL CARLSON, Ph.D.

OUR AWARENESS of the biological basis of our humanness is often confined to ignorance and tinged with fear. Our reluctance to educate ourselves and our children about the importance of the human condition stems from the fear that this knowledge is too disturbing and confronts us with our own mortality and imperfection. Yet it is the knowledge of this burden imposed on us by our biology that can liberate us from doing nothing to change our human condition. It can also inspire a respect and appreciation of ourselves that can help us cope with the tragic sense of life that this human condition conveys to so many of us.

We begin the human condition with the attempt to have children, a process that readily happens for most married couples but eludes some 20 percent after a year of effort to become parents. Half of these infertile couples will eventually conceive, often after consulting a specialist or fertility clinic for diagnosis and treatment.[1]

ELOF AXEL CARLSON, Ph D., is a Distinguished Teaching Professor, Department of Biology, State University of New York, Stony Brook. In 1965 he was the recipient of the Distinguished Teaching Award at the University of California, Los Angeles, and in 1972 he received the E. Harris Harbison Award for Distinguished Teaching from the Danforth Foundation. His interest in the field of genetics arose from his close association with the late Hermann J. Muller, Nobel prize winner at Indiana University. Carlson has published monographs concerned with the life and works of Professor Muller and has edited Muller's works. His teaching career included positions at Indiana University; Queens University, Kingston, Ontario; the University of California at Los Angeles; and the State University of New York, Stony Brook. He is a Fellow in the American Association for the Advancement of Science and a member of the Genetics Society of America, the History of Science Society, and the Honorary Society of the Sigma Xi. He is the author of *The Gene: A Critical History* (1966) and has publications in *Genetics, American Journal of Human Genetics, Quarterly Review of Biology, Nature, Journal of Theoretical Biology, Canadian Journal of Genetics and Cytology, Science,* and *Journal of College Science Teaching.*

Yet the other half remain childless for life, forced to settle for a lifestyle that neither partner expected, unless they were fortunate in obtaining an adopted child.

If the couples do succeed in achieving a pregnancy, the fetus still faces the risk of aborting, usually before the third month of pregnancy. About 20 percent of all women experience such a spontaneous abortion. Some 40 percent of aborted fetuses contain an abnormal number of chromosomes in the nuclei of their cells.[2] About 10 percent of all sperm or eggs will carry an abnormal chromosome number; this is a staggering fact to contemplate when we are so used to the textbook perfection of the biological concepts and principles we learn in our elementary courses.

If our fetus has passed this hurdle in its life cycle, it still faces the risk of being born with a birth defect. Some 5 percent of newborn babies have an impairment of sufficient seriousness to require medical attention.[3] In many instances this condition will not be curable and in some instances it is lethal, killing the child before it reaches adult status. Even where the birth defect is treatable, the emotional trauma to the parents and the financial obligations imposed on them may be staggering.

Are we finally out of the woods if the child is normal at birth and shows no signs of impairment during the first few years of life? No, for even now each individual must face life with a load of mutations derived from the unique shuffle and deal of genes the reproductive process passed on to it. Some individuals have unluckier combinations of mutations than others. The average load is about eight mutations, any one of which, if it had been passed on by both parents, would have been fatal.[4] Even in this buffered condition, where the potential damage of the mutation is compensated by a normal gene from the other parent, a slight impairment is felt. Higher genetic loads, if they do survive to maturity, often result in premature death, a need for surgery, or a weaker constitution subject to more frequent disease and medical care.

In the not too distant past the higher genetic loads did not live to maturity. The child born in 1900 in the United States had a 1 in 3 chance of dying of an infectious disease. Of the 10 major causes of death at the start of this century, 6 were infectious diseases. The number-one killer was pneumonia, about 12 percent of deaths; closely approaching this figure was tuberculosis; in third place with 8 percent of deaths was diarrhea of the newborn.[5] Children drank raw milk and sanitary conditions were primitive, even in our technologically advanced cities. Even such diseases as diphtheria and typhoid fever each accounted for 1 percent of all deaths then.

The infectious disease often made the children of a family all come down with a fever and other symptoms, but sometimes one child might die while the sisters and brothers would pull through. If, as some geneticists have inferred, the child that died had a higher genetic load than the survivors of the infection, then infectious disease played a major role in eliminating mutations from the population's gene pool and kept a balance between new mutations arising spontaneously and an equivalent amount being extinguished by the intense selection against the individuals with weaker constitutions.

We changed society dramatically in the twentieth century by introducing pasteurization of milk, sanitary handling of food, antiseptic hospital conditions with births in hospitals rather than at home, visiting nurse programs to instruct mothers on child-rearing practices, and compulsory immunization programs. Later, antibiotics were added to the strategy, which we devised to minimize life-threatening infectious diseases in our lives. Today some 70 percent of individuals die of heart disease, strokes, and cancer, and few infectious diseases remain on the list of the major causes of death.[6] This means that only in our own century have the technically advanced countries reduced natural selection in man and permitted an increase in genetic load.

The genetic load must increase in the absence of selective elimination because new gene mutations arise every generation. Each gene has only a slight probability of being mutated, about 1 in 100,000 times. This is a remarkably efficient process. If you compare a gene's accuracy to a good typist, the gene is by far a more efficient copier than the typist. A frequency of 1 typographical error for every 10 pages typed would delight most typists, but a gene's efficiency would be the equivalent of 1 typographical error in a manuscript copy of the complete works of William Shakespeare. Despite this remarkable efficiency, 10 percent of our sperm or eggs contain one new mutation that arose spontaneously. This startlingly high mutation frequency arises because each reproductive cell has at least 10,000 genes that constitute the instructions for the biological uniqueness of a new individual.

While we can concern ourselves with the prospect of an increased load of mutations in our population, much of the world is no different today from the days of our grandparents and our great grandparents. Children born in Chad, Burundi, and many other African countries still have a mean life expectancy of 20-30 years, a figure comparable to that of life in ancient Rome and not much better than our ancestors enjoyed in Philadelphia on Independence Day

1776.[7] We should recognize, however, that the mean length of life is determined largely by our environment. The biological length of life, by contrast, is fixed by our genes at some 75–80 years of age. This can be seen by comparing the mean age of life of philosophers and other intellectuals of the Golden Age of Greece (67 years) with that of comparable intellectuals today (e.g., Nobel laureates, Pulitzer Prize winners) who do not live appreciably longer than that.[8]

Similarly, if we compare mean life expectancy in the mid-nineteenth century United States with today, we would have a 30-year difference in a newborn child's prospects for life. But for individuals who were already 60 years old, the difference then and now in life expectancy is only about 3 years, although more than a century of medical progress has elapsed. The sharp depression in life expectancy is almost exclusively a consequence of childhood mortality. In countries where malnutrition accompanies inadequate public health programs, the mortality from such childhood diseases as gastroenteritis, measles, and whooping cough is high. Chile, for example, has nearly 1,600 deaths per 100,000 newborn in the first year of life from gastroenteritis, about 400 per 100,000 from measles, and about 70 per 100,000 from whooping cough. In countries with very efficient public health services, such as Sweden, the figures are dramatically reduced—only 13 per 100,000 from gastroenteritis and none at all from measles or whooping cough.[9]

The human condition has not traditionally been identified with an underlying genetic basis. Indeed, until the mid-1950s human genetics was a subject for medical curiosity of little practical value for the education of physicians and was generally held in disrepute.[10] This historical tradition stemmed from the rarity of individual genetic disorders, such as albinism, Huntington's chorea, hemophilia, or dwarfism, and the high incidence of childhood mortality for metabolic disorders such as cystic fibrosis, sickle-cell anemia, or Tay-Sachs syndrome. In most cases the genetic cause was masked by the immediate cause of death from pneumonia. Physicians were not too enthusiastic about concerning themselves with hopeless cases whose cause resided in a familial defect for which they had neither the knowledge nor power to intervene in its treatment or its reoccurrence. Genetic counseling did not exist before 1940. Even more dismaying were the spurious movements of eugenics, based on misconceptions of genetics and personal prejudice directed to the poor, the sick, the immigrant, the black, or the Jew. These attempts to preserve or improve the heredity of a population failed because they assumed wrongly that many social traits (economic success, criminal activity, mental illness, poverty, ghetto living) were largely genetic.

They also erred in believing that sterilization was the most efficient system for eliminating defective heredity from society. They were ignorant of the immense reservoir of heterozygotes whose mutations produced little noticeable impairment to them but provided the next generation with its birth defects when unlucky encounters of mutant genes took place. The more naive eugenicists assumed that these defects arose chiefly from adults who also expressed these traits.[11]

Many geneticists recognized the errors of these eugenic programs but did little to refute these mistakes and prejudices. The eugenics movement did not fail because these faults were exposed by experiment and careful scientific presentations; rather, the Great Depression of the 1930s created such widespread poverty that any claim of a genetic basis for "pauperism" was held up to ridicule.[12] Also, the unambiguous personal prejudice of Adolph Hitler and the Nazi movement he led made the literate world aware of the danger of identifying ethnic groups with superior or inferior heredities. Ironically, virtually no geneticists were in the vanguard of these movements attempting to apply genetics to man.

The revolution in human genetics began in 1956 when the human chromosome number was unambiguously shown to be 46.[13] The techniques developed by Tjio and Levan were soon applied to classes of institutionalized patients who had well-defined disorders that did not fit known genetic or environmental causes. A category of chromosome disorders involving the loss or gain of a chromosome was rapidly analyzed. Down's syndrome (called mongoloid idiocy in earlier days) was the first to be recognized as a departure from the normal 46 chromosomes.[14] Down's syndrome children have one extra member of their smallest chromosome. The hundreds of genes on that extra chromosome cause the cells to improperly turn on or turn off the various activities governed by their genes. Extra doses of enzymes and structural proteins are also likely to damage the tissues or organs of these patients. As a result, the newborn child with Down's syndrome has numerous physical and physiological defects. It is a semilethal disorder, imposing a biologically shortened life span (about 20 years) even with reasonably normal medical care. Most of its victims are retarded with IQs of less than 40. While a few may benefit from education to achieve a limited financial independence as assembly-line workers, porters, dishwashers, or baby-sitters, most do not, and they constitute a substantial part of the population of institutionalized retarded children whose expenses are borne by the state. Parents of such children are often faced with heavy medical expenses and few of them can afford private care if the child has too many problems to be managed at home. It costs the state about

$250,000 to care for a Down's syndrome child from admission to death. In the past these children seldom reached their adolescence because their immune system was so impaired that they succumbed to infections such as measles, pneumonia, or whooping cough.[15]

Several rare conditions, causing death shortly after birth, were also recognized for extra representatives of slightly larger chromosomes. In Down's syndrome there are three number 21 chromosomes represented, giving rise to the term trisomy 21 as a synonym for it. Similarly, there are Edward's syndrome (trisomy 18) and Patau's syndrome (trisomy 13), which have numerous skeletal, facial, and nervous disorders as well as deformities of the internal organs. By contrast, most of the defects involving the sex chromosomes permit the newborn child to live an impaired but otherwise nearly normal life span. These chromosome defects reveal an interesting, if unnamed, law of human heredity: The more biologically disastrous the genetic defect is to the fetus, the less socially significant it is to the parents or to society.

When a fertilized egg bears a defect so severe that the embryonic cell mass fails to implant in the uterus or aborts during the first few weeks after its attachment, the mother will be unaware that a conception has taken place. If the fetus has defects that cause its abortion during the second or third month, some parental sorrow may be present. If a miscarriage occurs after the mother and father can perceive its growth and witness its thumping and moving about, their grief is more pronounced. A stillborn child or a neonatal lethal is more tragic and may overpower the parents as they involve themselves in their child's struggle to live. A child born apparently normal, with a defect such as Tay-Sachs syndrome, will begin to show signs of its mutant gene during the sixth month of its life and get progressively worse, dying after 2 or 3 years of paralysis and convulsions. This catastrophe leaves a life-long sorrow and personal devastation to most parents of such a child. It may cause their marriage to fail; they may face bankruptcy for their medical expenses ($25,000-$100,000); and they may face psychological collapse from their inability to cope with their tragedy.

Parents do not have a philosophic belief in the equal worth of all conceptions. Their emotional response will be proportional to the investment of their hopes and familiarity with the growing fetus. This relation between genetic damage and social consequence forms an important part of the parents' response to genetic counseling.[16]

The study of human chromosomes was stimulated in the 1970s by the introduction of stains that can distinguish segments of the individual chromosomes. Depending on the degree of coiling and other properties of the chromosomes, dyes such as quinacrine or Giemsa can reveal a detailed banding that makes all 23 pairs of our chromosomes uniquely observable under a microscope.[17] The use of somewhat earlier stages of chromosome coiling now reveals about 2,000 different bands in our chromosomes. This detailed composition of the chromosomes can be used for identifying the breakpoints of exchanges between unrelated chromosomes; it reveals small duplications or deletions of genetic material, and it allows geneticists to see how certain segments are inverted or rotated 180 degrees. These defects are helpful in determining whether a disorder is familial and in predicting which members of the family are free from the defect. The cytogenetic revolution also permits geneticists to map genes on the chromosomes and to study the abnormal chromosomes often seen in tumors.

About the same time that the cytogenetics revolution began, a number of physicians and biochemists began a study of metabolic errors in man. These defects were first explored at the beginning of the century, but inadequate biochemical tools and the lack of a real understanding of how heredity works prevented a serious effort at working out any of these defects. During the 1940s and 1950s geneticists worked out the role of genes in synthesizing enzymes and other proteins. The molecular basis of mutation and its reflection in altered proteins was applied to man; the first case tied defects in hemoglobin to sickle-cell anemia and other related blood disorders. Still other genetic conditions were explored by a search for a defective enzyme or protein whose malfunction caused the symptoms of the genetic disease. In a few instances knowledge of the defect permits treatment. This is true for phenylketonuric idiocy (PKU), a rare disorder that can be partially prevented by early use (less than a month after birth) of a diet low in phenylalanine. The children with PKU cannot convert phenylalanine to tyrosine. The flooding of the cells with excess phenylalanine forces other genes that synthesize other amino acids to reduce their output. The nervous system then suffers from protein starvation and does not develop normally. When a proper amino acid diet is supplied to the infant, the nervous system develops normally or nearly so. Unfortunately, the brain damage in an infant treated a month or more after birth is irreversible and this forces physicians to use an early detection system, such as a mass screening program, to identify these children.[18]

The treatment of a disorder does not always follow the biochemical analysis of a gene mutation. The rare and pathetic paralysis of Tay-Sachs disease is caused by a defective enzyme, hexoseaminidase A, which in the normal brain cells metabolizes fatty substances deposited in nerve cells. When these are not properly recycled they are stored in large amounts. Although the Tay-Sachs child is symptom-free at birth, the nerve cells of such a baby already have 300 times the normal amount of fatty substance metabolized by hexoseaminidase A. No treatment is available to save such children. Only in a mere handful among some 2,000 known genetic disorders is a biochemical treatment available by use of altered diets or by supplying hormones and other biological products.

In a somewhat large sample, about 30-50 disorders, the defective enzyme can be detected by its failure to carry out a normal reaction in a test tube or other suitable biochemical screening test. Often these enzyme defects are present in the cells of the developing fetus, which float in the amniotic fluid of the mother's womb. Such cells can be sampled by a technique called amniocentesis. The test involves a mother who is about 16 weeks pregnant. Earlier attempts to withdraw fluid from the embryonic sac are too risky to the fetus. About 3 weeks of cell culture are required before an adequate test can be carried out. This means that the parents will not know if a child has a defect such as Tay-Sachs syndrome until the fetus is nearly 5 months old. This is a very advanced stage, often identified with movement and noticeable kicking. For such parents the choice is unpleasant — to request medical abortion of what they already believe to be a child or to allow it to be born and face the terror of seeing it die over the next few years.

Amniocentesis does detect, with virtually 100 percent accuracy, all gross chromosome abnormalities. For many women this is now an accepted procedure. The frequency of births with abnormal chromosome numbers increases with maternal age but remains constant in males. Thus mothers approaching age 40 have about a 1 percent risk of having such children. This is several times higher than those in their early years of reproduction. For such women, amniocentesis and medical abortion of the abnormal fetus is the only available medical procedure that can prevent the birth of a child with a chromosome defect. It is an impossible choice for many other women who reject medical abortion outright. Here, as in many areas of medical genetic practice, the moral and ethical conflicts are unresolved and both physicians and their patients have multiple and often contradictory attitudes on what is compatible with their consciences.

Amniocentesis does more than spot chromosome defects. It can be used for several metabolic disorders stemming from single gene defects such as Tay-Sachs syndrome or from multigenic defects of lower familial incidence such as spina bifida (which has a 3 percent chance of reoccurrence). In this neural tube defect, fluid from the spinal cord leaks out into the amniotic fluid. The alpha fetoprotein present in the sample indicates that the abnormality is present. As in the chromosome abnormalities, medical abortion or prevention of birth and not treatment of the defective fetus is available.

Amniocentesis raises many issues that go beyond the controversy of medical abortion itself. At first sight, the selective abortion of a child with Tay-Sachs syndrome might imply that the gene for this disorder will gradually disappear from the population. Actually, the practice of amniocentesis may increase the incidence of the gene because the child homozygous for the defect would have died anyway. But in the past such parents chose birth control, vasectomies, or tubal ligation rather than risk a repeat of the trauma such a child imposed. The inverse relation between the onset of severe biological damage and the emotional effect on the parents makes it more likely that the carriers of Tay-Sachs, having aborted a fetus with the disease, will try again. Mendel's law predicts that they have a 3 out of 4 chance of having a normal child. But these normal children have a 2 in 3 chance of being heterozygous carriers like their parents and would thus perpetuate the gene in the population. Neither the parent nor the physician today would seriously consider aborting a heterozygous child, particularly since this abortion, we should remember, will be applied in the fifth month of pregnancy. Another problem for parents and their physicians concerns the fetus with a relatively mild defect, or at least one that is still unresolved. Should a parent request that the sterile, Klinefelter syndrome eunuchoid be aborted or are his problems as an XXY male manageable?[19] Should a Turner's syndrome fetus, almost dwarflike in size, forever prepubescent and sterile, be allowed to go to term if the parents request that it be aborted? Would one choose the abortion of an XYY male whose abnormality, if present, is only an elevated risk of aggressively violent behavior and that attribute still much debated in medical circles?

All medical genetic practice is controversial because treatment permits gene defects to spread and add to our genetic load, but prevention raises the accusation that society practices eugenics and diminishes the variability of our characteristics. If we neither treat nor prevent we would be guilty of indifference to the victim of a genetic disorder or to the parents whose only wish is to have a normal child. It is my belief that both treatment and prevention are proper medical practices for genetic defects. The means to achieve these ends are limited. Direct treatment of metabolic disorders is available to a few dozen birth defects among thousands whose causes are still unexplored or remain enigmatic. Surgery can repair many organ defects from cleft palates to club feet, from blue babies to blocked guts. But many children have multiple birth defects arising from their genetic defect, and those affecting the brain, liver, or kidneys are more difficult to approach surgically.

The reality today is that for most birth defects little can be done. Genetic engineering is far removed from today's understanding or technology of molecular genetics. How would one go about localizing the many genes involved in a multigenic trait like anencephaly where the neural tube fails to develop into a full-sized brain? How would these be excised and replaced? If these miniscule acts of genetic surgery were somehow applied to a sperm, why would genetic engineering be preferable to selection of semen from another donor? So far there has not been a single case of genetic engineering successfully applied to the prevention or treatment of a human birth defect.[20]

Attempts have been made to use viruses as carriers of normal human genes with the hope that these will bypass the barriers our cells have to degrade foreign entering molecules. Perhaps someone will succeed. But this is a treatment, not a prevention. The target cells are few in number and even if some normal function is restored we know little of the long-range effects of such viral-carried genes. We do not know if malignancies will occur in response to such cell manipulation. Nor do we know how much normal function can be restored by the insertion of normal genes in a few cells. The prospects of altering every cell in an abnormal organ or tissue are remote.

Similarly, very few defects can be treated by enzyme therapy, where concentrations of the missing enzyme are injected into appropriate tissues. Few intracellular enzymes will pass through cell membranes and even those capable of functioning outside the cell may be engulfed as foreign invaders by our immune system. There is no doubt that replacement of enzymes and the products they make will be more commonly practiced in the decades to come, but for our generation, prevention, not treatment, is the more realistic choice for parents at risk to having children with birth defects or high genetic loads.

This leaves amniocentesis with its implied medical abortions as one choice. The second technique available, if adoption is unavailable, is even less popular—the choice of a semen source other than the father (and in rare instances the use of eggs other than the mother's). This procedure, called artificial insemination by donor (AID), is used by more than 10,000 parents yearly in the United States but most of them involve sterile males.[21] Although the use of AID is not intended to be eugenic, the selection procedure by the physician is somewhat eugenic. The donor, usually a fertile medical student, is carefully matched ethnically (as well as by blood group and by appearance) to match the sterile male. The physician usually asks about family illnesses and birth defects or other signs of severe medical problems. If the donor's sperm is acceptable to the physician it is then used anonymously so that the parent never knows the identity of the donor. Since students with psychoses, severe neuroses, or debilitating disorders that would interfere with their ability to pass a medical curriculum or to practice medicine are rarely admitted to medical schools, the semen derived from this source is likely to contain fewer defects than a sample chosen randomly from the population. Thus these conceptions should have the same advantages as children conceived and raised under rigorous natural selection. AID is not often used by a male with Huntington's chorea or one who is a carrier for a defect not detectable by amniocentesis. The availability of eggs is very limited and the technology of freezing and storing sex cells is limited to the considerably smaller sized and more numerous sperm. Frozen semen, thawed out after 10 years, has been used to fertilize eggs by AID. Two normal female infants have resulted from such long delayed use of the stored semen.[22] We still do not know what damage, if any, freezing and thawing may do to sperm.

Who imparts this knowledge of human genetics to the parents or concerned individual seeking advice on a genetic defect? In recent years the role of the genetic counselor has shifted from the geneticist (usually a Ph.D.) to the M.D. (specializing in medical genetics or pediatrics). In the past the counselor offered risk figures, explained pedigrees, discussed some of the problems the parents would face, but remained neutral in offering advice even to the point of refusing to give a personal opinion if asked. Today the M.D. counselor adds a much more detailed account of the medical problems and management of the child with a birth defect but still adheres to the policy of neutrality on advice.

This raises a social issue. Who takes the responsibility for the world's gene pool? If all counseling is directed to the family and if no attempt is made to inform the parents of their contribution to genetic load through their heterozygous carrier children, we must either deny that the gene pool gets polluted or assume unknown compensations that make human genetics immune from the consequences of increased genetic loads. I believe that a counselor can raise this issue with tact and concern, allowing parents to make their choice on the basis of family needs and their social conscience. The counselor is in the best position to do so because it is at this time that the relation of mutation to their personal lives can be extended to humanity itself; and their decision to have fewer children than the average, if amniocentesis and abortion are elected, could be interpreted by them, with the counselor's support, as a social good.

The number of individuals seeking genetic counseling is miniscule when compared to the staggering number of children born with birth defects each year. To change public attitudes substantially on the quality of life, I believe we need a major reform of the biology courses taught in high school and of the introductory course in college for majors or nonmajors. Biology can be taught effectively when it is tied to the human condition. To awaken our consciousness about the biology of our reproduction, our heredity, our character traits, and our racial differences, and to explore the controversies and ethical decisions these aspects of our life provoke, is a liberal arts tradition. By making our students aware of the mechanisms by which radiation and chemicals cause mutations, we make them concerned about the additives in their foods, the disposal and environmental contamination of radioactive nuclear materials, the proper attitudes to adopt in a doctor's or dentist's office when taking an x-ray, or the ways to change public policy. By studying the biology of cancer cells the student learns what is boxed in by the nearly cryptic warning that smoking is dangerous to one's health. A good deal of defensive living can arise from such courses, which can guide students in the habits that minimize their risks to themselves and their future progeny while permitting them choices they did not know exist. By studying the history of the eugenics movement or the current controversy on the heritability of IQ, the student realizes that science in the public sphere is subject to prejudice, distortion, error, and political manipulation. These are valuable lessons in a democracy. The aim of education should not be to hide decision making nor to avoid the presentation of controversy, but the reverse. It is by exploring the biological basis of these social issues that students learn to recognize pseudoscience, slanted presentations, and flawed theories with hidden assumptions.

There are many spurious issues in human genetics that divert attention from the problems that need to be faced. Such projections as cloning or the artificial production of hordes of identical twins have struck fear in the public mind that this is the direction of science applied to society. Aldous Huxley's *Brave New World*, we should remember, was a novel presenting no hope, for the world was as mad under a conditioned scientific totalitarianism as it was in the neglected, sickly, chaotic world of the savage. Cloning is of little interest to parents or to human geneticists. It is not identity but individuality that is preferred by society. Even mad dictators would have to wait some 20 years before their clones could serve as armies; and clones, like natural identical twins, will have separate identities. Furthermore, dictators never live as long as their "thousand year Reichs" and their mortality guarantees that new leadership will modify their policies and prejudices. From the geneticist's perspective a new generation produced by the union of sperm and eggs provides the combination of traits that can be sifted by human evolution and can lead to a subtle change in our capacities. This cannot be achieved by cloning, which perpetuates a constant genetic constitution repeating itself through time and diminishing the novelty of human evolution.[23]

Another spurious issue is the possibility that we will be replaced by chimeras composed of our brains and assorted artificial or unnatural organ systems. The specter of humanoids and synthetic humans is part of our science fiction heritage. It is not likely that we will do this to ourselves. The technology to achieve this and the individual cost to drastically alter our bodies to meet new functions are far beyond the capacities of contemporary science and society. Man has prevailed without substantial change in appearance and intelligence for at least 5,000 generations. It is not too unlikely that we can perpetuate our species for some 5,000 generations into the future with a minimum of compensatory differential breeding to maintain our genetic load at current levels. If we do direct our own evolution, it is my belief that these will be toward better health and lower genetic loads than toward unusual isolates reflecting the idiosyncracies of social planners. Only a small minority of humanity today will elect idealistic goals of higher intelligence or the behavior fostering human social cooperation as the traits they select for differential breeding. The fear that uniformity would replace diversity is unfounded. Neither AID nor individual family planning to foster these goals would have more than a minute effect on the processes of recombination that would shuffle these genes and guarantee the uniqueness of each individual conceived. Both intelligence and cooperative behavior are multigenic traits with major environmental components and in both cases the combinations for expressing these traits are numerous.

Human medical genetics is still an infant field of science. We know a lot about basic mechanisms but know very little about the specific defects involved in thousands of genetic disorders. We know very little about the functions of the individual genes in a multigenic disorder. We do not know how genes are turned on and off in human cells, whether normal, mutant, or malignant. By the end of this century we will know a lot more but our knowledge will still be applied to the treatment and prevention of only a portion of all known defects. Yet it is this increasing knowledge of our human condition that will give us the technology and value decisions we will make in our own behalf. It is the recognition and teaching of the biology of the human condition that can make us cope with our life-styles and the tensions in society. Whatever medical benefits human genetics provides, we can at least appreciate that it can give us a world view of our tragic and heroic features.

REFERENCES

1. Guttmacher, Alan. 1969. Birth Control and Love, pp. 235-68. New York: Macmillan.
2. Hirschhorn, Kurt. 1973. Chromosome Abnormalities. I. Autosomal Defects. In Medical Genetics, pp. 3-15. Edited by V. A. McKusick and R. Claiborne. New York: Hospital Practice Publ. Co.
3. Banister, Philip. 1972. Evaluation of Vital Record Usage for Congenital Anomaly Surveillance, pp. 119-35. In Monitoring Birth Defects and Environment: The Problem of Surveillance. Edited by E. B. Hook et al. New York: Academic Press.
4. Muller, H. J. 1950. Our Load of Mutations. Am. J. Hum. Genet. 2:111-76.
5. Taylor, Ian, and Knowelden, John. 1964. Principles of Epidemiology, pp. 20-51. Boston: Little, Brown and Co.
6. Grove, R. D. 1973. Vital Statistics. Britannica Book of the Year.
7. Dublin, Louis, and Lotka, A. 1935. Length of Life, p. 44. New York: Ronald Press.
8. Ibid., p. 32.
9. Bengoa, J. M. 1969. Malnutrition and Infectious Diseases: The Surviving Child. Biotechnol. Bioeng. Symp. 1:256.
10. Muller, H. J. 1949. Progress and Prospects in Human Genetics. Am. J. Hum. Genet. 1:1-18.
11. Ludmerer, Kenneth. 1972. Genetics and American Society. Baltimore: Johns Hopkins Univ. Press.
12. Muller, H. J. 1932. The Dominance of Economics over Eugenics. In A Decade of Progress in Eugenics: The 3rd International Congress in Eugenics, pp. 138-44. Baltimore: Williams & Wilkins Co.
13. Tjio, J. H., and Levan, A. 1956. The Chromosome Number of Man. Hereditas (Lund) 42:1.
14. Lejeune, J., Gautier, M., and Turpin, R. 1959. Etudes des Chromosomes Somatiques de Neuf Enfants Mongoliens. C. R. Acad. Sci. (Paris) 248: 1721.
15. Apgar, Virginia (ed.). 1970. Down's Syndrome (Mongolism). Ann. N.Y. Acad. Sci. 171:303-688.
16. Larson, Arlene. 1974. Counselling Revisited. Ph.D. dissertation, Univ. of Minnesota.
17. Caspersson, Torbjorn, and Zech, Love. 1973. Chromosome Identification by Fluorescence. In McKusick and Claiborne (eds.), Medical Genetics, pp. 27-38.
18. Hsia, David Y., and Holtzman, Neil. 1973. A Critical Evaluation of PKU Screening. In McKusick and Claiborne (eds.), Medical Genetics, pp. 237-45.
19. Several books have wrestled with these issues of bioethics. Among them are: Etzioni, Amitai. 1973. Genetic Fix. New York: Macmillan Publ. Co.; and Fletch-

er, Joseph. 1974. The Ethics of Genetic Control. New York: Anchor Press/ Doubleday.
20. The theory, however, is not lacking for such proposals. The most recent is the extension of gene insertion techniques from microbial systems to mammals or the reverse. The controversy on how dangerous this procedure is still remains unresolved. See Science 187(1975):931-35.
21. Guttmacher, Alan, Birth Control and Love, pp. 269-92.
22. Sherman, J. K. 1972. Unpublished letter to Dorothea Muller, July 1972. Muller Archives, Lilly Library, Indiana University.
23. Carlson, E. A. 1973. Eugenics Revisited: The Case for Germinal Choice. Stadler Genetics Symposia 5: 13-34.

DISCUSSION

Murray: Dr. Carlson, as a physician I feel it is incumbent on me to make some comment about the eugenic nature of using medical students as donors of sperm. Having been a medical student in the not too distant past, I would say that I would doubt there is any relationship between genetic vigor or lack of disease—even emotional disease—and medical students compared to the general population, or certainly the populations from which they are drawn. Getting into medical school is so much a function not only of how compulsive you are but of luck and being in the right place at the right time that I would wonder about your surmise that the selection process by which they are chosen is really some valid measure of their genetic contribution.

Carlson: I consider it slight but significant.

Moraczewski: If I recall correctly, Dr. Carlson, in one of your papers you mention that, taking lawyers, physicians, and Ph.D.'s as a group, the average of their IQs is about 130. Then you looked at the National Academy of Science and found their members had an average IQ of 165, an interesting differential. I would ask the question: Is there a difference between "raw" intelligence, which IQ measures more or less grossly, and creativity?

Carlson: Yes, there is a considerable difference between intelligence and creativity. There is a very interesting book by V. and M. Goertzel titled *Cradles of Eminence,* which studies about 400 different twentieth-century eminent individuals. Eminent people were defined as people who had two or more biographies written about them (who were not sports figures) and whose names appeared in the index file of the Claremont, New Jersey, Public Library. The mean IQ of these 400 individuals was 127, certainly not the genius category by Lewis

Terman's or by anybody else's definition. Furthermore, in contrast to Terman's California study where the high IQ types loved school, came from very stable middle-class homes, and were themselves very stable, the very creative individuals in the Goertzels' study usually despised their school experiences, and two-thirds of them came from troubled homes. So the Goertzels concluded that creativity is not measured by intelligence tests. Whatever term we apply to the loose word "genius," we have to remember it isn't always measured by standardized testing.

Walters: I have a question about the extent to which human genetic defects are preventable: If we had time and money to study the complete genetic makeup of every potential parent in the United States, what fraction of actually occurring birth defects could we predict on the basis of doing the complete genetic analysis of parents?

Carlson: For nonchromosomal genetic defects it would be a trivial number, considering the available knowledge we have today of birth defects. I doubt that it would have a major effect on the next generation but, if done regularly as our knowledge increases, it might make a difference on the genetic load of future generations, since the length of life of parents and grandparents is a rough index of what children themselves can expect.

Shaw: Dr. Carlson, you commented on siblings who came down with the same infectious disease, one carried off by death and the other surviving. Then you commented on the selection component. I was not clear whether there were genetic characters for immune defense for a particular disease or whether you were postulating that a generally genetically superior individual had survived the infection.

Carlson: I think there is more than immunity when it comes to survival after suffering a serious disease. I consider the whole genotype to be involved. Survival is a reflection of the genetic load, not just of the immune system.

Murray: Is there any evidence for this?

Carlson: No, it is an inference.

Murray: Father Moraczewski, I would like to ask about your formulation for characteristics. You went through some formulation by which you could look at heroes throughout history. I just wondered if this was a serious study, because of the kind of situation that might exist in a society full of heroes.

Moraczewski: It is something I concocted when I first came to the Department of Psychiatry at Baylor College of Medicine on a postdoctoral fellowship. I sat in on "grand rounds" where the staff meet for the purpose of discussing particular cases. One thing I noticed very quickly was the difficulty there was agreeing on what was normal. I started thinking about it. How could we empirically define or describe a normal person who would not be culturally determined? It occurred to me that possibly one way would be to search for clues in the epic literature. I chose epic literature because it represented the distillate of a particular culture at its best and (this perhaps is an assumption) that the heroes of such literature would represent what was noble and worthy in the estimation of that culture—the kind of person one would like to be. On that supposition, we could select the heroes from different epic literatures—Oriental, Near East, French, American, German, etc. (all the principal cultures of different times)—and identify those personality traits that were common, presuming there would be certain common traits. I realized there would be differences. For example, I noticed when reading Homer's Odyssey that the hero did not have the virtue of humility. Humility was not a characteristic of the Greek hero. Thus, establishing a hero-typology could be one way for empirically arriving at what may be called a transcultural universal noble man or hero—the "normal" man. What would you do in a world full of heroes? That is a good question. What do you think would be some of the problems?

Murray: I think they'd all be depressed because they would have nobody to rescue.

Carlson: I think the only characteristic found in common among mythical heroes is that they all have feet of clay.

Murray: I wonder, Dr. Carlson, about *your* formulation concerning the characteristics that you conceive as desirable.

Carlson: I don't know of any set of characteristics that all of humanity would agree upon. But people do make choices. When parents, some 20 years ago, were given a child for potential adoption, they had an inherent veto right. If the child had a serious birth defect, they could refuse to take the child. In many states there is still a certain time period after initial adoption procedures to permit the parents to return the child to the state if birth defects show up, such as Tay-Sachs disease. Many parents requested a specific sex as well as certain physical attributes in the child and these were usually honored. I remember talking to adoption agencies 15 years ago in the Los Angeles County area, when adopting parents could even reject children if one of the parents used LSD. That was the time when some spurious LSD fears were booming. I recognized that people in general made choices that were modest. They wanted a normal, healthy child. They weren't specifically looking for a genius or one with special talents. It occurred to me that if a married couple were sterile and there were no adoptable babies and, consequently, they had no choice but to use genetic sperm banks or artificial inovulation they would make relatively wise choices. You would not find an abundance of a few heroes even if there were available dossiers for all the geniuses, the ordinary people, and the crackpots who decide voluntarily to donate semen. I believe most people would choose what they felt most satisfactorily defines a normal child for them. I would promote the principle of voluntarism. If voluntarism does nothing else, at least it would prevent genetic deterioration from taking place. I don't have any hopes or any stronger desire that a positive eugenics system be the outcome of a sperm or egg bank program. I do worry about the genetic load and I feel this is something that should be a matter for parental concern.

Murray: I think there is a little contradiction between an experience in England and what you have just said. About 6-8 months ago, there was an article in the *New York Times* about the Ministry of Health in England deciding it would control the donation of certain kinds of sperm because there was an unusual run on requests for sperm donated by the leader of the Rolling Stones, Mick Jagger. The comment was that the ministry was not concerned about Mick's use of drugs or some of the other uncomplimentary things that were said about him but were concerned about the uniformity in the development of the gene pool. They must have had quite a number of requests. So I am not sure that the choices people make would be "reasonable," unless, of course, they see Mick Jagger as one of our epic heroes.

Carlson: There may be individual instances of adulation of a public figure. The reality is that most parents who elect artificial insemination by donor (AID) have their own personal tragedy. They are not coming for help because they want to be eugenic. In fact, they don't know anything about genetics. They just know they have a sterility problem. What they want is a child. The reality is that they are presently taking the physician's wisdom as the source of that child and have virtually no choice in it themselves. I am not sure this is the best way to do it.

Shaw: To get back to the epic hero. I'd like to ask Father Albert how he will square the sex ratio since most of our heroes are male.

Moraczewski: It would be interesting to determine the role of women in the past. We would probably find it much greater than we now realize. I don't think the literature actually reflects the powerful role that women have had in the past. It brings to mind a book by Joseph Campbell titled *Hero of a Thousand Faces,* which brings out the common traits of man. But the role of women has not been adequately discussed in the epic literature; it would be an interesting task. In terms of your question, I would say right now we would have to revise my program.

Walters: Dr. Carlson, in reading the literature by geneticists who discuss the problem of genetic load, there seem to be differences in assessing the significance of the genetic load and the degree to which it is currently increasing. In your view, what leads geneticists to differ in their assessments of the genetic load?

Carlson: There are two theories among geneticists concerning genetic load: The first assumes that newly arising mutations are found in carriers or heterozygotes. The mutant in this state may exert a very slight negative effect that eventually leads to its elimination under competitive conditions. This assumption is based on a series of experiments carried out with spontaneous and x-ray-induced mutations in fruit flies in the 1940s. Flies that carried any one of these new mutations in heterozygous condition did less well than flies that were homozygous for the normal gene.

The second school is the theory of overdominance, which assumes that there are certain genetic situations where a heterozygous individual is more fit than either the homozygous normal or the homozygous mutant. The best example of this in man is sickle-cell anemia. Actually, sickle-cell anemia illustrates the weakness of this overdominance theory because it is not an inherent superiority of the heterozygote. It is a heterozygous compromise, based on a kind of opportunism. The heterozygote state of impaired hemoglobin enabled children infected with malaria, especially the severe form of malaria that kills, to survive because such individuals were not as sick. Normal individuals do not survive because they usually die of malaria, and homozygous mutants do not survive because they usually die of anemia. This selection against homozygotes gave a balanced condition that always pumped the defective gene back into the population. When Africans came to the United States, that gene no longer provided a selective advantage because the malaria that is characteristic of equatorial areas in Africa does not exist here. As a result this gene is now at a selective disadvantage because the homozygous mutants die from anemia. Sickle-cell anemia in well-developed countries should lead to sickle-cell extinction unless medicine intervenes to save these anemic individuals so that they can reach reproductive maturity. In that case the gene would be maintained in the population. Geneticists have argued for more than 30-40 years about the experimental designs to test these two models and still do not agree on the degree to which each model contributes to our heredity. My own personal opinion is that, with the exception of some long-standing selected genes, virtually all newly induced individually isolated mutant genes that can be tested will turn out, whether induced or spontaneous, to have a genetic load effect.

Moraczewski: Dr. Carlson, just what is genetic load?

Carlson: Briefly, genetic load is the partial expression in hidden form of a mutant gene having a normal compensating gene from the other parent. As a result, that mutant gene does not directly kill the individual but reduces the effectiveness of that individual's function. If the gene produces an enzyme, that enzyme's activity is diminished, and the slight diminution of the ability of the organism to survive would be seen under severe test conditions.

Moraczewski: Another question frequently asked is: Is Down's syndrome an inherited disease?

Carlson: No, Down's syndrome or trisomy 21 is not usually an inherited disease, though occasionally it can be. For example, there are about 30 instances of females with Down's syndrome who have had children with this disease. In those instances, there is the possibility of their passing it on to one-half of their children. In a general sense, normal parents do not have the hereditary tendency to pass on Down's syndrome, but about 5 percent of parents with a Down's syndrome child may have a condition where a small part of their ovaries or testes contains a segment of cells all of which contain this extra chromosome. They have a much higher risk of having another child with Down's syndrome. That is why amniocentesis is advised for a parent who does have one such child. Dr. Shaw also pointed out that there is another category, called translocation Down's, in which case it is familial, because the number 21 chromosome is hooked up to another chromosome. The parent who has the abnormal hookup is himself or herself normal but the children have a very high risk of having Down's syndrome — about 30 percent.

Moraczewski: We've had several questions about amniocentesis, both involving the technique and why it is dangerous in early pregnancy and less so in later pregnancy. What is the possible damage to the fetus?

Carlson: The technique as it is practiced now on a 16-week-old fetus has very little risk. In 1975 the *New York Times* reported a survey of nine cooperating hospitals and over 1,000 attempts at amniocentesis and no significant difference in survival or malformation of the fetuses that went to term when compared with matched controls. Therefore, as it is practiced, it is a safe technique. The risks when it is done earlier include such difficulties as puncturing the fetus — the smaller the size of the amniotic sac the more difficult it becomes even when ultrasonics are used to locate the amniotic sac. Furthermore, enough fluid has to be withdrawn to permit culturing. The earlier this is done, the greater the relative amount of amniotic fluid that will have to be withdrawn, threatening the possibility of the fetus aborting. Those are hazards likely to be involved in doing amniocentesis earlier. Perhaps there will be medical advances where it can be done at the twelfth week. Perhaps Dr. Murray would know if there is research on attempts to do this safely at an earlier stage.

Murray: I recall that there were some early studies in amniocentesis done by Dr. Fuchs in 1955 showing that when he attempted to remove fluid as early as the twelfth week there was a higher frequency of spontaneous abortion following the procedure. The volume of amniotic fluid in the first trimester of pregnancy is so small that removing it changes the pressure relationships in the amniotic sac so drastically that it is likely to produce damage. The uterus at this time is much smaller and the likelihood of hitting the fetus or of missing the uterus altogether and hitting the mother's bladder or other structure is also greater.

Shaw: In 1975 it was reported at the American Society of Human

Genetics meeting in Baltimore that some people at Indiana University were developing a technique whereby they could obtain, by scraping, trophoblast cells that were sloughed from the fetus into the internal opening of the cervix. This does not require a needle and requires only an instrument like one would use to scrape the inside of the cheek, hardly anything more complicated than a Pap smear. They did obtain cells at 8 weeks of gestation and made direct smears to determine the sex of the fetus. They were 86 percent right in the number of fetuses that were aborted or went on to term. They did miss some males and in those cases felt they had maternal cells. They have now been able to grow the trophoblasts in culture. They grow much faster than the maternal cells and overgrow them, so perhaps we have the beginning of a new technique of prenatal diagnosis applicable in the first trimester.

Moraczewski: Can certain physical defects, such as harelip, web feet, or other defects of appendages, be discovered by amniocentesis, or perhaps fetoscopy?

Carlson: I'm hoping that the diagnosis of congenital defects will be increased by these techniques, but I don't know whether these techniques could get most of the multigenic defects of the 2,000 or more cases of the typical Mendelian characters. It is very difficult. Each one of these individual studies was the product of several years effort to work out an effective molecular or biochemical answer to what the defect is.

Moraczewski: Are most miscarriages due to genetic defect?

Carlson: Chromosomally, about 30–40 percent are. There are an unknown number of genetic defects in the strict Mendelian sense. We can't estimate the analysis of human genetic defects as readily as we can in other animals, such as fruit flies or other organisms used in experimental systems, because most of the induced or spontaneous mutations in man are eliminated in the embryonic state rather than occurring in the adult or live-born condition. So I suspect there is a significant number of miscarriages due to genetic defects — but what percent, we don't know.

Moraczewski: Why do some defects not show up until adulthood — for example, Huntington's disease?

Carlson: Because genes act at different times. They are usually in a turned-off state and require triggering by hormones or physiological changes in the body, so some of them may not show up until later in life. The classical example is pattern baldness in males. A teenage male does not know if he will be bald or not from pattern baldness until there is a sufficient amount of male hormone causing failure of hair growth. Exactly why certain diseases like Huntington's chorea require 30-50 years before they are manifested may have to do with the accumulation of products slowly building up and reaching a point where they impair function. That is something I think could be studied at different stages by biochemical or molecular studies of tissues of individuals who have Huntington's chorea.

Moraczewski: In regard to Huntington's disease, what are the chances of a child of such a parent having the condition?

Carlson: Fifty percent with each pregnancy, something Arlo Guthrie will soon discover for himself since he has married and decided to have children. If he is also affected, he will have the same 50 percent risk for his children as Woodie Guthrie had for him.

Moraczewski: In regard to AID, if parents really want children and they are not able to have them the normal way, why not adopt them rather than using such involved processes as AID?

Carlson: This is the consequence of a very important social phenomenon. In New York a white couple wishing to adopt a white child will have an average wait of 4-5 years for an adoptable child. Since New York State also imposes a matching of religion between child and adopting parents, the wait for certain groups (Jews, for instance) may be as long as 10-20 years, because there are so few available Jewish children for adoption. In older days adoptions were easier because there was a stigma attached to women who had a child out of wedlock and they would frequently abandon such children because they would not be able to function with the personal shame they felt for their error. Society has changed. It has changed since Margaret Sanger's family planning, birth control, and legalized abortion have reduced the number of children born out of wedlock. It has changed because of the massive amount of contraceptive techniques developed during the last 10 years that greatly reduced the number of unwanted children. Many women who have a child out of wedlock today take an entirely different attitude. They don't consider their child as stigmatized by their family or by society. All these things have combined to make adoption an extremely difficult experience for individuals. Most available adoptions are for interracial children. There is some reluctance on the part of each race to adopt an interracial child as readily as they would a child of the same race. But even this attitude is disappearing in the population. By the end of the century I would be very surprised if adoption constituted more than a very minor solution to childless marriages. The genetic reality will remain the same, 10 percent of these couples are going to be sterile. They will have to choose: Do they wish to be childless and live that way, or do they wish to try something else? And the only available procedure now is the slim possibility of adopting a child or having a child from a gamete source—egg or sperm. In today's technology the only practical reality is from donated sperm. It isn't male chauvinism that determines sperm banks, it's the abundance of sperm and the difficulty of obtaining and storing eggs.

Moraczewski: We have a general question here that is a fundamental one and appears in a number of formulations. I remember when I first ran across the idea. Dan Callahan, a few years ago, said we were moving toward a society where we are going to produce children in a quality control environment such as directs the production of automobiles on the assembly line. Children will have to meet certain specifications, otherwise they would be eliminated. Such an attitude of perfection would seem to militate against the reality of mankind. When a child is born who is not perfect as defined by IQ, health, and other tolerance standards, the parents feel guilty that they have not produced a perfect child. Hence you seem to be promoting a goal of perfection, where all genes would be without any defects, all persons would have a high IQ, would be nonaggressive, or minimally aggressive, and so forth. Is that eugenics, is that the kind of program geneticists would like to promote? Would they want the elimination of all defective genes and the promotion of some genes that would be agreed on by society?

Shaw: I would like to broaden that question a bit and not limit it to genetics. I think we are seeing people with higher expectations in other areas of life, such as owning two cars, or a TV, or having sanitation, or not expecting a third or fourth of their children to die of infectious disease during the first year, and then feeling that they are cheated if this is denied to them. I think our culture is imposing very high expectations on us and on our life-styles in many different ways, and this desire for the perfect child is one aspect of a much larger issue.

Carlson: I don't think it's the desire for a perfect or ideal child. I believe it is the wish of expectant parents to have a normal child. They do not welcome the idea that the child will pathetically waste away despite all their efforts or what medicine can do for it. They don't want a child so impaired that the child is incapable of having a social life after the parents are gone. But if the real desire is normalcy, which I believe is the case, then I feel it is a false issue to claim that geneticists are somehow promoting a positive genetic program as their leitmotif to life. While many geneticists, for example, H. J. Muller under whom I studied, did believe this and considered it part of their credo, most geneticists do not; and I think it is a mistake to take this idealism, hold it up as an impossibility, set it up as a straw man, shoot it down, and put blinders on our eyes and say mutation doesn't matter; let any and all mutants pass into the population; the more there are, the more variety for mankind. But what is this variation we want? Does it include every biological catastrophe that can happen? Do we want an unlimited amount of damage to keep building up over the next thousands of generations of man? Who would put on the brakes? Where are the repairs coming from? Would we just keep on patching ourselves to the point where almost all our resources and all our ingenuity go into self-repair? Or do we want some standard of normalcy without our imposing heroic proportions? I believe that it is normalcy we strive for, and by using that term I do not mean mediocrity, but the reduction of known mutations and higher genetic levels.

Murray: Two highly loaded terms have been used: one, perfection, and the other, normalcy. And I would venture to say we could have a symposium lasting for weeks or months just discussing the meaning of those terms. I remember from Herman Muller's last lecture, a swan song, at the International Congress of Human Genetics in Chicago, that the only characteristic he, after all the years of thinking about this problem, agreed would be desirable for the human race was general cooperativeness, because then we would perhaps not destroy one another in our hopes and desire to improve ourselves. So I would say we would perhaps forget about the term "perfection" and try to get some idea, not of normalcy, but what is acceptable in the species that allows us to cooperate and work with one another, or what other kind of variability we would accept within those limits. What we would look for would be the limitations to place on what would be acceptable for the human population. But even that is a tough decision.

Walters: I think we are going to have to develop two kinds of attitude simultaneously: One is the effort to find some means to prevent the occurrences of genetic defects—prevent increasing the genetic load; on the other hand, and at the same time, we have to cultivate attitudes of tolerance toward those who are genetically defective. I think it is going to be a very difficult task to achieve these two goals simultaneously but we are going to have to attempt it.

2/
Genetic Counseling: Boon or Bane?

ROBERT F. MURRAY, JR., M.D.

GENETIC COUNSELING! What is it, who does it, and what good is it? It is a special type of medical advising in which the genetic counselor provides information and education but, unlike the doctor, tries to avoid giving advice. It may sound crazy but it isn't when the process and philosophy of counseling are understood. What I mean can be illustrated if genetic counseling is contrasted with medical advising as it is usually practiced.

There is little question that in medical counseling the doctor gives the patient advice. It is unusual for an M.D. to truly leave the decision up to the patient even though it might, on the surface, seem that way. For example, where a decision to have surgery for an ulcer is involved, the physician might review possible courses of action and then leave the decision to have or not have surgery up to the patient, but it is usually made quite clear to the patient what the doctor wishes him to do. The patient might be told something like this:

ROBERT F. MURRAY, JR., M.D., is Professor of Pediatrics and Medicine and Chief of the division of Medical Genetics in the Department of Pediatrics at Howard University, College of Medicine, Washington, D.C. He is a Fellow of the American College of Physicians; a Fellow of the American Association for the Advancement of Science; and a Fellow and a member of the Board of Directors of the Institute of Society, Ethics, and the Life Sciences. He was a member of the Social Issues Committee of the American Society of Human Genetics. He has served on several committees of the National Research Council of the National Academy of Sciences, including Ad Hoc Committee on Sickle Cell Trait in the Armed Forces, Chairman; Committee for the Study of Inborn Errors of Metabolism; Committee for a Study for Evaluation of Testing for Cystic Fibrosis; and Committee on Maternal and Child Health Research. He has been elected to the Institute of Medicine of the National Academy of Sciences. His publications are in the *Annals of Human Genetics, Nature, British Medical Journal, Science, Annals of the New York Academy of Science,* and *Ghana Medical Journal.* The author of a number of monographs in genetic studies, he has contributed to proceedings and compilations of essays in the field of genetics.

It looks like your duodenal ulcer, the most common type of ulcer, isn't healing the way it should. We could continue the therapy for a while longer and see what happens, but in my experience, when an ulcer hasn't healed by this time, it isn't going to. Besides, there might be a hemorrhage or, if the ulcer gets worse, it might perforate the duodenum (the first part of the small intestine) and cause peritonitis, which is a serious life-threatening condition. The surgery itself isn't very risky. We remove $\frac{1}{3}$ to $\frac{1}{2}$ of your stomach and cut the nerves to the stomach. The chances are that you won't have any more trouble with that ulcer. After an initial period of adjustment, most of the people who have surgery are able to eat what they wish without difficulty. If you don't have the operation, the ulcer might heal but there is always the chance that it will recur and you will have to go through the same problems as now. Of course, the decision to have the surgery is up to you.

It should be obvious that it would be very difficult if not impossible for a patient who believes in and trusts the doctor to choose not to have surgery after the situation is presented this way. If this were being presented according to the principles of modern genetic counseling, the risk at each step of the procedure would be presented and the odds for and against a favorable outcome of the surgical as well as the nonsurgical approach to therapy for duodenal ulcer would be presented and the meanings discussed. There isn't time to present a detailed list of the risks involved in the choice of surgery for duodenal ulcer but some of the major factors to be considered include:

1) The risk of injury and/or death from anesthesia.

2) The risk of postoperative complications, e.g., adhesions might develop between adjacent parts of the intestine and lead to obstruction.

3) The risk of a poor or inadequate result of surgery, e.g., the occurrence of an ulcer at the place where the stomach and intestine are joined together.

4) The risk that the remaining stomach may not adapt or stretch so that its function does not normalize and the patient has continued inadequate nutrition.

These risks also ought to be balanced against:

1) The probability that the ulcer will heal without complications.

2) The chance that the ulcer will not recur.

From a consideration of these risks, a composite probability of failure might be calculated and also a composite probability of success would be determined and presented to the patient who would then decide whether to choose surgery or to choose to continue medical therapy. According to one source, the risk that complications may result from medical therapy are 26 percent while 20 percent of patients have an unsatisfactory result.[1] In 5 percent of this group, the surgical complications are worse than the original ulcer. With this kind of information, many persons might prefer not to have surgery.

As I describe this process, I realize that it sounds a great deal like the procedure one might go through to get true informed consent. But there is a significant difference in genetic counseling because one vital outcome in the cases where genetic counseling counts is a *reproductive decision* that is to be made by the individuals, usually couples, involved in counseling. The physician or counselor may or may not be involved in the action taken based on the decision whereas, in the usual medical model of decision making, the doctor will almost always be involved. In other words, genetic counseling can be considered a process of informed decision making. To reproduce or not to reproduce, that is often the question!

WHO WANTS GENETIC COUNSELING?

To gain a better understanding of genetic counseling, it might help to get a picture of who gets counseled or seeks counseling.

Most of the time, a couple who has produced a child with one or myriad birth defects that may or may not have a genetic basis seeks an answer to the question, Will this happen again and, if so, is there anything we can do to avoid the birth of another such child? Not infrequently, the couple is actually at the end of their reproductive years and is more interested in knowing what can be done to treat the disorder or take care of the affected child. A prime example of this situation occurs in Down's syndrome (popularly known as mongolism) where the risk of a child with this condition being born to a woman over 35 is 10-45 times greater than to a mother in her early 20s.[2] In fact, the available evidence suggests that most congenital malformations are more likely to occur in older mothers who are, as a rule, less interested in having additional children.[3]

Much less frequently, the couple who both carry the same deleterious gene that determines an autosomal recessive or non-sex-linked trait and who have with each pregnancy a 25 percent risk of having a child with a genetic disease have been identified in a screening program. The couple want information about the condition since they often have had no first-hand experience with it. They want to know how it is inherited and what they can do to avoid having an affected child. As screening programs for more common genetic disorders like sickle-cell anemia and Tay-Sachs disease become more widely accepted and other inherited diseases such as cystic fibrosis can be screened for, couples with this kind of problem will appear much more frequently.

In our clinic and others, a significant number of single persons who have been identified as carriers of deleterious genes are counseled. These persons, usually teenagers or young unmarried adults, have been sent or seek information about the medical and social significance of the carrier condition and what they can do about it. One small but very important group of persons who seek counseling may carry an unusual or translocated chromosome that is "balanced" in their own genetic makeup but is responsible for a moderate-to-high risk of recurrence (about 10–30 percent) of Down's syndrome in the offspring of the carrier parent (usually the mother).

People seek or are referred for counseling for a variety of other reasons. Sometimes they just want information. Not infrequently, they learn things that can have a drastic, sometimes devastating effect on the course of their lives. Such might be the case of the person who learns that an uncle recently died of Huntington's disease — a chronic, progressive, deterioration of the nervous system — a dominant trait that is caused by the inheritance of a single abnormal gene. Persons who have inherited the gene have usually had their children before they themselves show any signs of it. The often unexpected revelation that there is a significant risk that the person who has come seeking genetic information is not only at risk but that any children they have will also be at risk can produce an abnormal degree of anxiety, depression, or both. Humane counseling requires the counselor to be supportive not only medically but also psychotherapeutically. Counseling is a very comprehensive process that involves giving information, education, psychotherapy, and sometimes medical therapy. The genetic counselor who gives information that upsets the course of the person's life is obligated to provide support while the consultant reorients that life.

A committee of the American Society of Human Genetics has summed up the comprehensive nature of humane genetic counseling in its recommended definition:

> Genetic counseling is a communication process which deals with the human problems associated with the occurrence of a

genetic disorder in a family. This process involves an attempt by one or more appropriately trained persons to help the individual or family to:
1) comprehend the medical facts, including the diagnosis, the probable course of the disorder, and the available management;
2) appreciate the way heredity contributes to the disorder and the risk of recurrence in specific relatives;
3) understand the alternatives for dealing with the risk of recurrence;
4) choose the course of action which seems appropriate to them in view of their risk, their family goals and their ethical and religious standards, and to act in accordance with the decision; and
5) to make the best possible adjustment to the disorder in an affected family member and/or to the risk of recurrence of that disorder.[4]

Where medicine is concerned with all diseases, genetic counseling is focused on the ones where abnormalities of the genetic material are presumed to be in large measure responsible. Where medical counseling is concerned with the life of the living patient, genetic counseling tends to emphasize the life of the unborn. Where the individual is of primary concern in medical advising, the family often takes precedence in genetic counseling.

HOW DID GENETIC COUNSELING AND GENETIC COUNSELORS EVOLVE?

Genetic counseling was not always so comprehensive nor did it have recognized medical status. In the 1920s and early 1930s, eugeneticists motivated by ideas from geneticists like Sir Francis Galton began to advocate programs of genetic hygiene or genetic advising with the hope of promoting the improvement of biological and social characteristics of people.[5] Sir Francis is known for his studies of hereditary characters in humans, using twins. Geneticists at the Eugenics Society Record Office carried out limited counseling for many years. In 1934 a symposium was held on genetic counseling that promoted the idea of physicians participating in the practice of negative eugenics through genetic counseling. Negative eugenics is the process of improving the genetic quality of the species by preventing mutant genes from being passed on to the next generation, thereby supposedly improving the quality of the gene pool. Although the first heredity clinic based at a university was established at the

University of Michigan in 1940, the Dight Institute for Human Genetics, which began operations in 1941, was the first agency in the United States to publicly offer genetic counseling services. Dr. Charles Dight, a physician, left the funds in his will to establish the institute named for him "To Promote Biological Race Betterment — Betterment in Human Brain Structure and Mental Endowment and therefore, in Behavior."[6] Despite the eugenic orientation of this mandate and the fact that the records of the Eugenics Society office were transferred there, the early consultations were more the basis for future studies than to help families requesting genetic information. Benefits to society as a whole were not involved in assisting these families. Dr. Sheldon Reed who wrote the first book on genetic counseling, *Counseling in Medical Genetics,* published in 1955, suggested the term genetic counseling in an early publication of the Dight Institute.[6] By 1951 there were 10 genetic counseling centers in the United States. In the past 25 years, that number has grown to 387, as listed in the 1974 International Directory of Genetic Services, and there are 890 units worldwide providing a great variety of services.

Many, if not most, of the professionals first involved in genetic counseling were Ph.D. geneticists who may have been strongly influenced by eugenic considerations in their counseling. They did not diagnose medical illness but provided risk figures and interpreted them. As the frequency of children affected by genetically determined disorders increased, medical genetics attained the status of a recognized medical discipline with the establishment of Divisions of Medical Genetics at the University of Washington, Johns Hopkins School of Medicine, and the University of Wisconsin in successive years. This, coupled with a rapid expansion of research in genetics, was instrumental in promoting a rapid increase in the involvement of physicians in genetic counseling. A recent nationwide survey by Dr. James Sorenson revealed that 80 percent of the genetic counselors in this country are M.D.'s, 40 percent of whom also hold the Ph.D. degree.[7] Most of them (63 percent) are pediatricians. Eleven percent of the counselors hold the Ph.D. degree. It is not surprising that these counselors adhere primarily to a medical model of counseling which is usually centered on the needs of the individual.

The need for counseling services has expanded dramatically in the past four years, not only because physicians diagnose genetically determined conditions more frequently and are more aware of the need for genetic counseling in inherited disorders but also because new and widespread programs of screening for inherited conditions like sickle-cell anemia and Tay-Sachs disease have been established across the country.

GENETIC COUNSELING: BOON OR BANE?

To meet this need, paramedical persons such as nurses, social workers, and health educators have been trained to provide counseling under certain special conditions. There are also three new programs established to train persons at the Master's level to provide primary counseling in hospitals in university settings. These counselors work as part of a team in most cases to deliver counseling for a variety of inherited disorders. It is quite likely that they may soon become the predominant group of counselors.

Within professional ranks, there has been a rather heated debate over the question, Who is best suited to counsel? Most persons working in the field agree that a team of individuals is required since it is unusual for one person to have all the skills needed. But who should be the primary person on the team?

WHO SHOULD DO GENETIC COUNSELING?

It might seem logical that the primary physician should counsel, since he is responsible for making the diagnosis but the following make him unsuitable:

1) He is not accustomed to spending the time (45 minutes to 1 hour) needed to provide comprehensive counseling, and this much time would also be very expensive.

2) He finds it difficult to be nondirective because he is a product of the medical model that emphasizes giving advice.

3) He is accustomed to doing most of the talking and is often a poor listener.

4) He may be insensitive to psychological cues.

5) He is often unfamiliar with genetic concepts and finds it difficult to communicate these to the lay person.

Not only have most currently practicing physicians had little or no genetics training but also they have a confused idea of the relationship between the significance of the carrier status and the disease, and almost 40 percent have little or no personal experience with genetic disease. An equal number have difficulty communicating genetic concepts.[8]

Ph.D. counselors lack the diagnostic skill required in the genetic workup since an accurate diagnosis of the condition involved is the first requirement for proper genetic counseling. There is some evidence from Dr. Sorenson's survey that Ph.D. counselors are better able to be nondirective, i.e., they are less directive than M.D.'s.[7]

Properly trained paramedical counselors or Masters-level counselors who have special training in communication skills and are better able to empathize with the laymen should be an improvement over the often poor communication skills of the physician. These professionals should also be able to provide the time for psychological support at a reasonable cost. It is probable that this latter group will become the primary counselors as the need for genetic counseling increases, because they can be trained more rapidly. Physicians will be responsible for making an accurate diagnosis, providing medical management in those cases where it is needed, and delivering genetic counseling in more complicated cases. It is possible that there will be significant changes in the orientation and improved communication skills of the primary physician as new developments in medical curricula and delivery of health care occur. These might make the physician better suited for the special interpersonal skills involved in genetic counseling.

IS GENETIC COUNSELING A BOON OR A BANE?

I have purposely avoided using the terms *good* and *bad* in discussing genetic counseling because the terms are so "loaded" that one cannot help but get trapped in an ethical quagmire. The contrasting terms *boon* or *bane* seem much less "slippery" in an ethical sense. *Webster's Third New International Dictionary* describes a boon as a "benefit or favor; one that is specifically asked for or is given as the result of a request; an often timely and gratuitous benefit received and enjoyed; a blessing." A bane can be "any pernicious or fatal element, feature, or flaw; curse; a person who makes another completely miserable; one that perversely or persistently thwarts."[9]

Genetic counseling can be both boon and bane but which it is or how much it is of each depends on the disease in question and the situation in which it is applied. It can be both a blessing and a curse; it can bring happiness or misery. Genetic information can make life seem bright and full of promise or empty and hopeless. The family may be presented with what appear to be insoluble conflicts that must be resolved through a tortuous process of weighing the medical, genetic, and ethical consequences of different decisions in which the consequences of any one will be undesirable. The following specific cases illustrate the kinds of conflicts that arise.

Tay-Sachs disease is an inherited degenerative disease of the central nervous system most frequently found in approximately 1 in 3,600 Jews with Eastern European ancestry at birth. It is an autosomal recessive trait, which means that each parent of an affected child carries one gene determining the disease but is himself unaffected by it. The diseased child has two such genes, having inherited one from each parent. The Tay-Sachs child is born healthy but at 3-6 months of age begins to deteriorate mentally so that he is left blind and functioning at a vegetative level until death occurs, usually by 4 years of age. About 70 percent of parents who have such a child have no further offspring, so devastating is the effect of this child on their psyche. The cause of this disease is a deficiency of an enzyme, hexoseaminidase A, which is vital in the degradation of complex compounds called gangliosides, which accumulate in the brain in excessive amounts. Fortunately, the biochemical defect can be identified in cells found in fluid removed from the amniotic sac that surrounds the baby by a process called amniocentesis. Thus the presence of a fetus destined to have this disorder can be detected at 14-16 weeks of pregnancy, early enough so that parents who wish can elect to have a therapeutic abortion. To such parents, if they have no ethical conflicts about abortion, genetic counseling is a boon because they can avoid the previous suffering they experienced and also avoid the 25 percent or 1 in 4 risk of bringing into the world a child who will "suffer" as their first one did. But counseling might be a bane if therapeutic abortion is unacceptable to one or both parents. This conflict is further heightened in the case where both husband and wife are identified as carriers of the gene determining Tay-Sachs disease as part of a screening program before they have had any children. Genetic counseling followed by amniocentesis can provide them with a way to avoid the 1 in 4 risk of having children with the disease. The destruction of the potential life of the Tay-Sachs fetus, who will die soon after birth, by therapeutic abortion is balanced by the opportunity to have children without the disease. The net effect may be life rather than death.

Programs of screening, counseling, amniocentesis, and therapeutic abortion carried out in high risk populations appear to be cost effective as well.[10]

Down's syndrome, better known as *mongolism,* is a disorder in which the affected child has a characteristic facial appearance, is mentally retarded to a varying degree, and frequently suffers from malformations of the internal organs and other abnormalities that significantly shorten the life expectancy. Many of these patients, most often born of mothers over 35, are living fuller and more active lives until middle age and beyond because of advances in medical care, rehabilitative services, and surgical therapy. Dr. Jerome Lejeune found that this abnormality was caused by an extra number 21 chromosome so that there are three instead of the expected 2 number 21 chromosomes.[11] This chromosomal finding is called trisomy 21. This and all other chromosome abnormalities can be detected in cultured amniotic fluid cells so that mothers 35 and older, at increased risk to have such children, can avoid their birth if they choose. But this decision is complicated by other considerations:

1) Many of the children with this condition have sufficient intelligence to be trained to do simple tasks and take care of themselves.

2) A majority of them are capable of affection and meaningful human relationships.

3) But in contrast to this is the fact that some who live into adulthood (an increasing number do) have no place to go and are a burden on their older parents, since they are usually born of older mothers.

To many persons, it is not obvious that children with Down's syndrome should not be allowed to be born because, despite the medical and social problems they create, they do have "some human value."

There are rare situations where the use of amniocentesis may not provide all the answers needed and so may result in a serious dilemma.

A 26-year-old female carrier of a 14/21 chromosome translocation experienced a period of moderate depression lasting 3 months following the birth and soon afterward the death of an infant with Down's syndrome. At 16 weeks of pregnancy, she underwent amniocentesis to identify the chromosome makeup of her infant because of the 1 in 5 risk of having another child with this problem. The result was reported as normal.

At about 7½ months of pregnancy, her obstetrician detected and confirmed the presence of twins. The parents were initially unaware of the diagnosis. At this stage in pregnancy, it is not possible to safely terminate the pregnancy and one cannot be certain of getting amniotic fluid from each amniotic sac separately so there is a 1 in 5 chance that one of the twins will have the 14/21 translocation and Down's syndrome.

There are several alternative courses of action open to the physician, no one of which is totally satisfactory:

1) He might tell the parents, in the interest of truth-telling and honesty. If the other twin is all right, things will be fine, but if the twin has Down's syndrome, the mother may have another serious postpartum depression.

2) He might not tell the parents. If one twin has Down's syn-

drome, the mother could be prepared before being told or, in an extreme case, the affected child could be held back and put up for adoption since the mother did not want it anyway.

3) The parents could be told and be given an explanation of the risk—that chances of the second child having Down's syndrome are approximately 2 in 15 or 13.3 percent. Emotional support should then be provided for the parents, especially the mother, through the remainder of pregnancy to help avert the postpartum depression that might occur.

Where parents get the answer they want or have a way to avoid the tragedy, genetic counseling and amniocentesis is a boon. But where unexpected and uncontrollable results occur, as in the previous case, it is both boon and bane and in some more bane than boon. For example, what of the case where amniocentesis is performed for Down's syndrome and an unexpected chromosome abnormality that is less serious is found, i.e., with no mental retardation or serious internal or external deformity (e.g., short stature in Turner's syndrome, poor sexual development in Turner's or Klinefelter's syndrome, or even worse, the XYY chromosome makeup with all the unanswered questions about its association with criminal behavior). How can parents who are prepared for one kind of news cope with something totally unexpected? Certainly, this is both boon and bane. The outcome of the pregnancy described earlier was a happy one. A normal nonidentical twin boy was born. The parents were told about the presence of twins after suitable psychiatric preparation. Continued emotional support through the latter stages of pregnancy was also provided.

BOON vs. BANE IN OTHER GENETIC DISEASES

X-linked Disorders

The Lesch-Nyhan syndrome is a tragic neurological disease inherited as an X-linked recessive trait. This means that mothers carry the gene responsible and each of their sons has a 50:50 chance of inheriting the gene and expressing it. These boys are severely retarded, have uncontrolled writhing movements of the arms and legs, mutilate themselves by biting off fingertips and chewing off pieces of the lower lip. They do experience pain, and, although they cannot help themselves, are aware of what they are doing to themselves. They are happy to have their arms restrained to prevent them from harming themselves. No treatment is known.

It is possible to diagnose this condition which is caused by a deficiency of the enzyme hypoxanthine-guanine phosphoribosyl transferase, which can be demonstrated in cultured amniotic fluid cells obtained by amniocentesis at the 14th–16th week of gestation. It is, therefore, possible for parents at risk to have such a child to avoid its birth when diagnosed by having therapeutic abortion performed. This would most certainly constitute prevention of suffering and, if the parents accept abortion, can be considered a boon for such families.

Duchenne's muscular dystrophy is an X-linked disease affecting males and is characterized by the progressive degeneration of skeletal muscle beginning in childhood and progressing to involve chest and heart muscles, which results in death usually before young adulthood is reached. Unlike the Lesch-Nyhan syndrome, this condition cannot be specifically diagnosed by examination of cultured amniotic fluid cells. But since only male offspring of female carriers will be affected and since it is possible to diagnose the sex of offspring by chromosome study of amniotic fluid cells, parents might opt for therapeutic abortion of male fetuses. Unfortunately, half of those aborted male fetuses will be unaffected with the condition. This would appear to be a mixed blessing; a boon is that parents can avoid having an affected child but a bane since to do so one must sacrifice some otherwise healthy male fetuses.

Hemophilia A is an inherited abnormality of blood clotting inherited as an X-linked trait. Until the recent development of a concentrate of clotting factor VIII, which corrects the clotting deficiency, boys with this condition rarely reached adulthood. Most died from internal hemorrhage caused by trauma. The factor VIII concentrate must be administered regularly to keep clotting normal but it requires large amounts of blood products and about $10,000 per year per person. Because of the expense, some states have passed laws authorizing the use of state funds to support the costs of administering factor VIII concentrate to victims of hemophilia. This boon for medical science, however, produces another problem. All the daughters of males with hemophilia will be carriers of the gene and in turn half the sons of these females will have hemophilia. The frequency of the gene and the condition determined by it will increase rapidly in the population. The need for this expensive treatment will increase and in time could become a significant drain on financial and medical resources. Is this a boon or a bane? Take your choice.

These cases involve situations where counseling offers some kind of alternative action albeit unpleasant in some instances where it might be considered more boon than bane. There are other common

GENETIC COUNSELING: BOON OR BANE?

conditions where the options available to the couples who must make a decision are, from certain points of view, negative ones — where counseling might seem like more bane than boon.

Sickle-Cell Anemia

Sickle-cell anemia is an inherited disease of the red blood cells caused by a minute change in the structure of hemoglobin. Hemoglobin is the protein that fills the red blood cells and makes them red; its chief function is to carry oxygen. Both sexes are affected by sickle-cell anemia with equal frequency. This is a disease whose cause has been known for 20 years but for which a cure has not yet been discovered. In the past 4 years, intense scientific, public, and governmental activity has been directed at this disorder that may affect 1 in 600 Afro-Americans at birth and is sometimes found in other ethnic groups. The rediscovery of this neglected disease led to poorly conceived public education and massive screening campaigns that not only detected children with the disease but also identified persons who were carriers of the gene that determines this autosomal recessive trait. What kind of counseling situations arise with this disorder?

1) Where one member of a couple is a carrier and the other is not, there is a boon, since the couple can be informed that they do not have to worry about having affected children.

2) Where both members of the couple are carriers with a 1 in 4 or 25 percent risk of having a child with sickle-cell disease there are both bane and boon — since they know they are at risk to have an affected child but have no positive alternatives.

3) A child is found to be a carrier of sickle-cell gene. Parents are tested, only one is a carrier. This is a boon because the couple are given the good news that they do not have to worry.

4) A child is found to be a carrier of the sickle-cell gene. Parents are tested and both are carriers. This is both bane and boon for the reason stated in case 2.

5) A child is found to be a carrier of the sickle-cell gene. Parents are tested and neither is a carrier of the gene. This is a bane because this situation raises the spectre of nonpaternity, which can threaten the stability of any marriage and seriously affect the self-image of the child.

The only documented attempt to deliberately eliminate sickle-cell disease through screening and reproductive counseling met with quite unexpected results.[12]

THE TRAGEDY OF ORCHOMENOS. Orchomenos is a small agricultural community about 150 kilometers to the north of Athens, Greece. The educational level of the people there is quite low since half have only an elementary school education and 10 percent are illiterate. Most marriages (⅔) are arranged by the families. One percent of newborn infants have sickle-cell disease and 23 percent have sickle-cell trait.

The entire population of this community underwent sickle-cell screening and counseling over a period of 3½ years. Ninety percent of those screened received genetic counseling in which they were instructed that persons both of whom carried sickle-cell trait should not marry since children with sickle-cell anemia might be born.

Seven years later, the program was reevaluated. Despite the fact that before marriage persons inquired about the carrier status of prospective spouses, there was no significant change in the frequency with which matings occurred between persons both of whom had sickle-cell trait. There was also no change in the frequency of children with sickle-cell anemia. But there were social and psychological changes that occurred as a result of this program. Persons with sickle-cell trait, especially females, experienced social stigmatization. This is not surprising since 25 percent of the families felt that having sickle-cell trait meant restriction of marital freedom and social stigmatization. Furthermore, 40 percent of normal individuals and 20 percent of carriers felt that sickle-cell trait was a mild disease even though they had been counseled to the contrary. The degree of stigmatization is strongly evident in the fact that parents in this community now instruct their children with hemoglobin AS (HbAS, sickle-cell trait hemoglobin) to avoid persons with HbAS. But further that 20 percent of parents who are HbAS instruct their hemoglobin AA (HbAA, normal hemoglobin) children to avoid marrying a person with HbAS. As a consequence of the social embarrassment of the carrier state, many females with HbAS traveled to other villages to seek mates where the significance of HbAS was unknown. Not only did this screening program fail to achieve its goals of preventing mating between sickle-cell trait individuals but it distorted the sociocultural roles of individuals in the community and introduced psychological stigmatization not previously present.

The term "sickle-cell trait" to designate the carrier of the sickle-cell gene is unfortunate since the term "trait" suggests a taint of the disease sickle-cell anemia. Eliminating its use and substituting sickle-cell gene carrier might help reduce and help eliminate some, and perhaps eventually most, of the stigmatization of the carrier status. It should be noted that the term "carrier" is used for all other genetic conditions.

The situation in genetic counseling in cystic fibrosis—an inherited, chronic, incurable disease affecting 1 in 1,500-2,000 persons of European ancestry, leading to death in young adulthood, and much less frequently affecting Blacks and Orientals—is similar to that in sickle-cell anemia. Unfortunately, there is only one test that will reliably and unambiguously detect the patient with the disease. This is the so-called sweat test—a complicated procedure requiring considerable skill and experience to perform accurately.[13] In its current form, it is not adaptable as a screening method. All other methods recommended for cystic fibrosis testing have serious flaws or sources of error. Even if a simple test for cystic fibrosis were available, the fact that there is no effective treatment for this disorder and no effective means for prenatal diagnosis means that, like sickle-cell anemia, the ratio of boon to bane is unfavorable.

It is only a matter of a few years or less before reliable methods of prenatal diagnosis become available for these latter two conditions. There is already a reliable, although technically difficult, method of making the diagnosis of sickle-cell anemia in the 14-16-week-old fetus. The only stumbling block remaining is to develop a safe method of getting red blood cells from the fetus or the fetal aspect of the placenta. Work on the technology to accomplish this is going on at several medical centers. The technique has already been used successfully to diagnose the absence of another blood disorder of hemoglobin synthesis, thalassemia major or Cooley's anemia, in a fetus at risk for the disease.[14] Methods that permit the diagnosis of less severe disorders raise complex ethical dilemmas, if not for families, for society, since they raise questions of human value. How much abnormality or deviation from "normal" are we willing to tolerate? Will parents be advised by future counselors to avoid the birth and possible baneful existence of any child not normal by standards set by large insurance companies or HEW? One might envision the development of a boon-to-bane ratio by which fetal fitness to be born would be judged. The ratio would be calculated by computer but the problem of establishing measurable parameters is a big one. This kind of estimate is already made in crude fashion by individual families when they consider the "burden" of giving birth and rearing such a child. Dr. Edmond Murphy has presented a hypothetical scale on which the burden of a disorder in arbitrary units is related to the recurrence risk.[15] This gives a measure of something he calls "expected load." The rather unpleasant and perhaps frightening thought about all this is the possibility, perhaps even the likelihood, that even when specific and effective treatment has been perfected for many of these disorders, the option to terminate life will continue to be chosen for eugenic reasons, for economic reasons, or for reasons of convenience. But there are genetic reasons why genetic counseling has limited applicability as a negative eugenic tool:

1) Our knowledge of the composition of the gene pool is inadequate and our understanding of the meaning of the unexpectedly large amount of genetic variability that has been found is grossly deficient. For example, it is possible that the sickle-cell gene has other properties that make it beneficial in addition to providing resistance to falciparum malaria in carriers.

2) To have any significant impact on the frequency of what appear to be deleterious genes, we would have to embark not only on massive compulsory programs of screening but also sterilization and/or abortion of persons who carry mutant genes. This would clearly be unethical and is grossly inconsistent with the current ethical standards that guide medical practice.

WHAT WILL GENETIC COUNSELING BE LIKE FOR THE TRICENTENNIAL PEOPLE?

Will genetic counseling be the same or different for the people of the year 2076? If limits are placed on family size to achieve zero population growth, and other resources we now take for granted are limited, what will be the attitude of the tricentennial person toward the acceptable quality of human life? Will there be compulsory screening programs at birth for hundreds of genetic traits with prohibitions against marriage between possessors of genes determining those traits mandated by the state? Or will there be compulsory genetic counseling for every couple seeking to be married? A new government agency may have to be established—the Department of Reproduction and Gene Control—which will stringently regulate mating so that persons with the same deleterious genes do not have children. This is, of course, an almost impossible task and could be achieved only if all reproduction occurred via *in vitro* fertilization.

Some geneticists have predicted that amniocentesis will become a routine part of the prenatal care in every pregnancy. Abortion of fetuses that do not meet certain health standards might then become mandatory. There would have to be a "new" kind of genetic counselor paid by the state to see that persons with "defects" were not born so they would not be a drain on its resources. There would be no concern for the needs of parents—only for the cost/benefit ratio or boon/bane ratio projected for the individual.

I want no part of such a program for it cannot help but end in promoting genetic conformity in the same way that there are and have always been pressures to promote cultural conformity. As a physician-geneticist, I feel compelled and believe it wiser to continue to steer the course we have steered in the past, namely, to meet the needs of the family and the individual fetus or child. I would prefer to see man become extinct in the process of following principles based on love and humane concern for the needs of our brother and sister human beings than to ensure our survival under regimented, inhuman programs in which we are programmed like so many computer punch cards.

About 99 percent of all the species that have existed on this planet are now thought to be extinct. Just as the physical end of each person is predestined at birth by the genetic defects in his DNA, so the end of the human species as we know it may be built into the composition of the gene pool. While we stretch our minds to find ways to rid our society of genetically caused disease, deterioration of our social and governmental systems may set the stage for a nuclear holocaust that may not only destroy human life but most other life on this planet as well.

REFERENCES

1. Hendrix, T. R. 1968. Abdominal Pain. In *The Principles and Practice of Medicine*, 17th ed., pp. 934-35. Edited by A. M. Harvey, L. E. Cluff, R. I. Johns, A. H. Owens, D. Rabinowitz, and R. R. Ross. New York: Appleton-Century-Crofts Educational Division, Meredith Corp.
2. Penrose, L. S., and Smith, G. F. 1966. Down's Anomaly. London: J. and A. Churchill.
3. Nora, J. J., and Fraser, F. C. 1973. Medical Genetics: Principles and Practice. Philadelphia: Lea and Febiger.
4. Fraser, F. C. 1974. Genetic Counseling. Am. J. Hum. Genet. 26:636-59.
5. Ludmerer, K. M. 1972. Genetics and American Society: A Historical Appraisal. Baltimore: Johns Hopkins University Press.
6. Reed, S. E. 1974. A Short History of Genetic Counseling. Soc. Biol. 21:332-39.
7. Sorenson, J. R. 1973. Counselors: Self Portrait. Genet. Couns. 1:31.
8. Rosenstock, I. M.; Childs, B.; and Simopoulos, A. P. 1975. Genetic Screening: A Study of the Knowledge and Attitudes of Physicians. Washington, D.C.: National Academy of Sciences.
9. Gove, P. B. (ed.). 1968. Webster's Third New International Dictionary. Chicago: W. Benton, Publisher, Encyclopaedia Britannica, Inc.
10. Childs, B. 1975. Genetic Screening: Programs, Principles and Research. Washington, D.C.: Committee for the Study of Inborn Errors of Metabolism. Division of Medical Sciences NRC, National Academy of Sciences.
11. Lejeune, J.; Gautier, M.; and Turpin, R. 1959. Etudes des Chromosomes Somatiques de Neuf Enfants Mongoliens. C. R. Acad. Sci. (Paris) 248:1721-22.
12. Stamatoyannopoulos, G. 1974. Problems of Screening and Counseling in the Hemoglobinopathies. In Birth Defects. Edited by A. G. Motulsky and W. Lenz.

Proceedings of the Fourth International Conference, Exerpta Medica, Amsterdam.
13. Howell, Doris A. 1976. Evaluation of Testing for Cystic Fibrosis. Report of the Committee for a Study for Evaluation of Testing for Cystic Fibrosis. J. Pediatr. 88:711-50.
14. Kan, Y. W.; Golbus, M. S.; Klein, Phyllis; and Dozy, André M. 1975. Prenatal Diagnosis in a Pregnancy at Risk for Homozygous B-Thalassemia. New Eng. J. Med. 292:1096-98.
15. Murphy, E. A. 1973. Probabilities in Genetic Counseling. In Contemporary Genetic Counseling. Birth Defects Original Art. Ser. IX: 19-33. White Plains, N.Y.: The National Foundation, March of Dimes.

DISCUSSION

Carlson: Dr. Murray mentioned that he teaches his course in human genetics at 8:00 a.m., and not long ago he taught it at 8:00 a.m. on Saturday, an impossible hour for medical students and himself. The reason for this is that the status of medical genetics in medical schools is still low, and I'd like to ask Dr. Murray why he feels this attitude still exists among physicians and what he feels can be done to change it.

Murray: I think the primary reason that medical genetics doesn't have status—and I hate to say this in a public gathering, but I think it is true—is that you can't make money at it. It is not a money-making venture like the specialties of surgery, radiology, or clinical pathology that generate money for medical schools or hospitals. In most states Blue Cross-Blue Shield will not pay for genetic counseling or for some genetic diagnoses. I know that Medicare in Washington, D.C., does not pay for it and we have to provide funds. Fortunately, our program is supported by federal grants, and if people can't afford to pay the usual fees, we see them and provide services free of charge. I think the economic factor is one of the primary reasons for the low status of genetic counseling. Another reason is that it is only recently that genetics has been introduced into the medical school curriculum and most physicians who are practicing have not been introduced to genetics and are ignorant of the value that it has in medical practice.

Walters: You have concentrated on the needs of the family in genetic counseling. Dr. Carlson has mentioned the possibility of suggesting to the family what its obligation to the gene pool might be. Do you think that an appropriate topic to bring up with the family would be the effect on the gene pool of the decision they might make, or do

GENETIC COUNSELING: BOON OR BANE?

you feel that would be a coercive kind of factor to bring into the counseling situation?

Murray: I believe it might be somewhat coercive for the socially minded individual but I think it's important for most of us to appreciate that there is still some argument among geneticists about the nature of the gene pool and the genetic variability in the gene pool. I feel it would be premature for me as a counselor to say to a family that they should be concerned about the kind of gene and the gene pool when we geneticists cannot be sure of the significance. Second, when one is counseling in the medical model, the assumption of the patient who comes for counseling is that you are going to have his welfare or the family's welfare at heart and I think if someone is going to raise the family's sensitivity of obligations to future generations then one should hang a sign on the door or hand the patient some information saying, "My primary concern is for future generations and not for your welfare and I counsel in that light." That way the people know what the counselor is thinking. It is certainly possible for the counselor, just as it is for a doctor, to play with the statistics in such a way that, even when the risk is very low, people will be reluctant to take the chance to try to have another child. For example, the risk of recurrence of enancephaly may be one in a hundred — a low genetic risk — but if you tell the couple that the risk for them is twenty-five times greater than people in the general population, that will probably have a deterrent effect. I think it should be made clear to counselees if a counselor is going to have as his primary goal in counseling a concern for future generations.

Moraczewski: Also, in that same situation you mentioned, there is a 99 percent chance of normalcy. There would be one chance out of a hundred to have the abnormal trait, and 99 chances to have it normal as far as that particular trait is concerned.

3/

Genetics and the Law

MARGERY W. SHAW, M.D., J.D.

WHY ARE WE hearing so much about genetic disease these days? Isn't it a problem that has always been with us? Yes, of course, it is, but our awareness is quite recent. There are at least two reasons for this. As infectious diseases have been brought under better control by drugs and immunization programs, genetic diseases have emerged as a major problem. Also, as our knowledge and understanding increase, many conditions are, for the first time, being recognized as genetic in origin.

Over 2,000 genetic conditions have now been identified, and for the past decade about 100 new ones have been discovered each year. It has been estimated that more than 90 percent remain to be discovered. Let me enumerate a few of the commonly known genetic conditions. These can be divided into three groups: single gene defects, chromosomal disorders, and polygenic disease. In the first category, you are probably familiar with cystic fibrosis, PKU,

---

MARGERY W. SHAW, M.D., J.D., is Director of the Medical Genetics Center and Professor of Medical Jurisprudence at the Health Sciences Center, University of Texas, Houston. She has received the Billings Silver Medal from the American Medical Association, the Achievement Award from the American Association of University Women, and the American Jurisprudence Award. She has been a Visiting Scholar at the Yale Law School and a Visiting Professor of Medical Genetics at the Yale Medical School. Her other teaching experiences include an associate professorship at the University of Michigan Medical School and a professorship at the University of Texas Graduate School of Biomedical Science. She has been a member of the Board of Trustees of the Tissue Culture Association, the Council of the Environmental Mutagen Society, the Medical Advisory Board of the Houston Planned Parenthood Association, and the National Genetics Foundation. She has publications in numerous journals, including the *American Journal of Human Genetics*, the *Houston Law Review*, *Science*, the *American Journal of Law and Medicine*, the *North Carolina Law Review*, and she has collaborated in other publications on genetic counseling and genetics and the law.

NOTE: The research for this chapter was supported by Medical Genetics Center Grant GM-19513.

hemophilia, muscular dystrophy, Huntington's chorea, sickle-cell anemia, and Tay-Sachs disease. The chromosomal disorders may be caused by extra chromosomes such as mongolism or Down's syndrome; missing chromosomes, as in Turner's syndrome or XO syndrome; and structurally abnormal chromosomes such as found in the cri-du-chat syndrome. The polygenic diseases are common conditions in which both genetic factors and environmental components play a part, as in diabetes, hypertension, hypercholesterolemia, schizophrenia, certain forms of cancer, and some cases of arthritis, blindness, deafness, and mental retardation. All our organ systems are susceptible.

What is the impact of genetic disease on the health of our society? It has been estimated that 40 percent of infant mortality is occasioned by a genetic problem, 25 percent of admissions to children's hospitals are due to genetic illness, and perhaps 30 percent or more of all the institutionalized handicapped are there for genetic reasons. More than 10 percent of us carry a hereditary enzyme deficiency that may make us hypersensitive to commonly used drugs.

Here are some additional statistics: More than one-third of all spontaneous abortions are caused by gross chromosomal defects, and 5 percent of all liveborn infants have a birth defect.

Genetic diseases are *chronic* diseases, present throughout the entire life of the affected individuals. We have treatments for some, but no cures. The emotional and financial burdens never let up.

If this picture is not bleak enough, then consider the fact that every one of us carries 5-10 defective genes that may be passed on to our offspring.

With this general background on the impact of genetic disease in our society, let us turn now to some of the problems raised by our new discoveries.

Many of the problems arise on a very personal level in a genetic counseling setting. Genetic counseling has been defined as follows:

> It is a communication process concerning the risk of occurrence of a genetic disorder in a family. It involves an attempt to help the person or family comprehend the medical facts, appreciate the hereditary nature and recurrence risks in specific relatives, understand the options for dealing with the risk, choose the most appropriate course of action, and make the best possible adjustment.

This is a tall order, fraught with many ethical and legal problems.

Let us examine the basic *process*—communication between counselor and counselee. The interaction between these two participants will, of course, vary with the personalities, cultural and educational backgrounds, attitudes, beliefs, and value systems. We might analogize the system to a physician-patient relationship even though the counselor is often not a physician and the counselee is often not a patient.

The law has delineated certain rights and duties in such a fiduciary relationship. In general terms, where there is an expert and a layman, the law places duties on the expert and grants rights to the layman. An example of this may be seen in the informed consent doctrine: the doctor has a duty to inform; the patient has a right to decide. In the genetic counseling context, then, the counselor must disclose the risks of recurrence of genetic disease in a manner that the counselee can comprehend. He must also disclose to the counselee the various options available to minimize or prevent recurrence. These options fall in the area of reproductive choices: abstinence, contraceptives, sterilization, abortion, artificial insemination, and adoption.

The duty to disclose is probably greater in genetic counseling than in a medical setting for two reasons. In medicine the physician is granted some latitude in disclosure, rightfully or wrongfully, under the dictum of the therapeutic exception. Where anxiety in the patient may interfere with his therapy, the doctor's judgment has been recognized by the law. However, there usually is no imminent treatment in the genetic scenario and thus latitude in disclosure fades. The second reason for complete truth-telling is founded on more philosophical grounds. The ethic of self-determination and privacy of family life place a clear primacy on decision making with those individuals who must live with its consequences.

This standard of complete disclosure has been argued by genetic counselors under the heading of directive vs. nondirective counseling. Some counselors feel that their duty is ended when they disclose the risks—they attempt to avoid involvement in the decision-making process. But disclosure of alternatives is just as important, even though some of the alternatives may be morally distasteful to the counselor. The counselee has the right to choose the best alternative for himself or herself. For the counselor to sidestep certain issues may be just as improper as imposing personal bias.

A particularly knotty problem in truth-telling arises in the not infrequent situation where tests to determine the genotypes of parents and offspring uncover the fact of illegitimacy. Should the counselor disclose his findings to the married couple he is counseling about future risks? If only the mother is a carrier of a recessive gene and an affected child is born, this couple has no risk in future children. Surely, they have a right to know. But would complete disclosure put a strain on the marriage? There is no easy solution to this problem. But perhaps the best course to follow is to disclose to the couple, *before genetic testing*, that such tests may uncover nonpaternity. Forewarned, the couple may choose not to be tested, or may waive their right to know the results of the tests. The same approach, i.e., pretest explanation of the possibility of unexpected findings, might also be used when chromosomal screening is done for whatever reason, in the event that an XYY may be inadvertently discovered.

To return to the problem of transmission and receipt of information in genetic counseling, we may properly ask the question whether the messages given about genetic risks are accurately comprehended. A number of follow-up studies have disclosed that only a minority of those counseled remembered what they were told about recurrence risks in future children.

Perhaps the most important factor in the counselee's interpretation of risks depends on the nature of the outcome if the gamble is lost. Thus the burden of the disease in question may influence the decision more than the mathematical odds.

In one reported study, Hsia and Silverberg found that as many as 75 percent of couples running a risk higher than 1 in 10 either decided to have more children or were still undecided. These investigators suspect that counseling helps to reinforce the innate desires of parents who want more children but is much less effective in influencing parents to decide against childbearing.

One wonders whether the counselors were so intent on allaying anxiety and providing reassurance that they subconsciously painted an optimistic picture. The manner in which the mathematical odds are presented may influence the way individuals perceive a risk. For example, a 1 in 4 risk of abnormality might be interpreted differently than a 3 to 1 chance of a normal child, although they are, of course, the same odds.

What about dissemination of genetic information to third parties? This issue arises in two contexts: first, in the case of relatives of the counselee who may be found to be at increased risk by virtue of their placement in the family pedigree; second, by more distant relatives or kindreds who may be discovered through linkage in a centralized genetic registry data bank.

The legal issues in these cases are those of confidentiality and the right of privacy. Among several meanings of the right of privacy is the freedom from public disclosure of private facts. Conceivably, close relatives or even distant relatives may not be classified as "the public." Of course, the counselor should utilize discretion and limit his disclosure to only those relatives who may be at risk. From a public policy perspective, disclosure to protect the health of a third party seems reasonable. Case law does suggest that the counselor may communicate medical information to the spouse of a counselee. Perhaps the best course for the counselor to take would be to obtain prior consent that if information is discovered that might be of importance to relatives it can be communicated to those persons at risk.

The counselor usually asks, "Do I have a *right* to contact relatives at risk?" Perhaps a better question would be, "Do I have a *duty* to contact relatives at risk?" In other words, when is it legally permissible or legally required to break the implied contract of confidentiality in order to warn others of dangers facing them? The Supreme Court of California [*Tarasoff* v. *Regents of University of California*, California Supreme Cts. (July 1, 1976)] held that the psychiatrist had a duty to warn the parents of a girlfriend of his patient, after the patient had revealed in confidence that he planned to murder the girl. No warning was given and the threat was carried out. Arguments in court centered around the psychiatrist's inability to predict violence and the necessity for complete confidentiality in the psychiatrist-patient relationship. In an analogy to genetic conditions, the risks are often accurately predictable and failure to warn might someday be construed by the courts as a dereliction of duty on the part of the counselor.

Schmickel has commented that the image of a geneticist is rapidly changing from a "bookie" to a "fixer." This is certainly true in the area of prenatal diagnosis and selective abortion of affected fetuses. Using the technique of amniocentesis during the fourth month of pregnancy, more than 80 genetic diseases and all gross chromosomal abnormalities may be detected. Thus certain parents may now be provided with the option, hitherto unavailable, to choose to bear only children who pass the diagnostic screen.

Amniocentesis is the procedure of withdrawal of fluid from the uterus by placing a needle through the mother's abdominal wall. The fluid contains cells sloughed off from the fetal skin and respiratory and urinary tracts, some of which are still living. These cells may be examined directly or grown in culture for chemical and chromosomal tests. In expert hands, the procedure is really quite safe for both mother and fetus, and it is no longer considered experimental. But some methods of prenatal diagnosis require the use of ultrasound, the direct observation of the fetal parts with a lighted instrument called a fetoscope, and withdrawal of a sample of blood from the placenta or the fetus itself. These latter procedures are in the experimental stages of perfection but may soon become routine.

Since amniocentesis must be performed during the second trimester and there is often 2 or 3 weeks delay in obtaining laboratory results, the whole problem of legalized abortion faces geneticists. This is particularly true since Dr. Edelin's conviction of manslaughter in Boston for a second trimester abortion. This has exerted a chilling effect on all obstetricians contemplating abortions after amniocentesis in mid-pregnancies. Accurate fetal diagnosis during the first trimester of pregnancy would be a welcome technological discovery.

This brings us to the whole area of legal issues in abortion. I would like to outline briefly some of the ways our courts have examined the legal rights of the fetus.

The first cases to receive judicial notice involved property rights of the fetus rather than personal rights. As early as the seventeenth century in England a living being *in utero* was recognized by the courts to take under a will, inherit an estate, or receive property under the laws of descent. This applied to posthumous children (children who were conceived but not yet born when the father died). Of course, these rights did not ripen until birth, but the estates were put into escrow for the potential human being.

These precedents on property cases carried over into tort law. A tort is a harm done by one human being to another and the usual remedy is a money damages award in a judicial attempt to make the person whole. A number of cases on wrongful death ensued. In these cases the court first recognized a being in existence when the fetus was viable in the latter months of pregnancy if an injury had occurred to the mother which harmed the fetus, but again these rights did not ripen unless a live birth ensued. In other words, if an automobile accident occurred that caused a stillbirth, no damages were collected on behalf of the dead fetus. Later the courts recognized quickening as the time of potential recovery and in 1946 the first case to recognize an early embryo injury resulting in death after live birth occurred. Of course, fetal injuries resulting in defects but no death were also recognized.

Another concept in law is the tort of wrongful life. In two cases where the mother had rubella during early pregnancy, suit was brought against the doctors and the hospital who refused to perform an abortion in the face of grave risks to the fetus. The baby did not recover in either case but the parents did recover in one case for financial loss and emotional pain and suffering. The courts agreed that they could not compare a damaged life against no life at all. These cases occurred before the Supreme Court decision in 1973 that legalized abortions during the first 6 months of pregnancy. This year in Texas another rubella case was decided in favor of the parents of the defective child. Damages were awarded for the extra medical costs only.

Traditionally, primitive societies have not recognized the fetus as a person but welcomed the live-born infant into the human community with all the rights and privileges thereto. The *Roe* v. *Wade* case, decided by the Supreme Court, also did not recognize a fetus as a person under the Fourteenth Amendment clauses of due process and equal protection. It is interesting to note that the Supreme Court avoided the issue of viability but merely announced that the states *may* proscribe abortion during the last trimester except to preserve the mother's life or health. They did not say *must*. Thus each state, by legislative action, has the option of legalizing elective abortion up until the time of birth.

Another interesting problem has arisen in regard to welfare payments and food stamps. Is the pregnant woman to be regarded as one person or two persons for the purposes of welfare benefits? Different states have interpreted the federal social security statute in different ways. Thus a California woman who was pregnant won the right to collect double food stamps. The United States Supreme Court ruled in 1975 that states are not required to provide extra benefits to pregnant women at any stage of pregnancy, again underlying their position that a fetus is not a person in the legal sense.

Congress and the states may not pass laws to prohibit abortion since the Supreme Court has spoken. The only way a change in the law would come about is by constitutional amendment or by the Court reversing its own decision on a later case. Some states have attempted to require consent of the husband or the parents of a minor child to interrupt a pregnancy. The Supreme Court has invalidated these laws on the principle of the right of privacy of the mother in deciding what she shall do with her own body.

Underlying these discussions on genetic problems is the issue of society's interest in monitoring and improving the quality of the human gene pool. What about legal restrictions on childbearing? Certainly the law recognizes an interest in the health of its citizens by passing many statutes and imposing many regulations for the express purpose of controlling infectious disease. Does this power of the state run to genetic disease regulation as well? It may be illuminating to compare the similarities between infectious and genetic diseases.

First, both are transmitted from one individual to another. In infectious disease, transmission is horizontal, through the present generation; in genetic disease, transmission is vertical from one generation to the next.

Second, both infectious and genetic diseases vary in their contagion rate. Some infectious diseases are highly contagious, such as Asian flu, with a predictable number of exposed individuals contracting the disease. The contagion rate or risk of genetic disease is also often predictable and, in some situations, very high. For example, in the case of an isochromosome or a reciprocal translocation between homologous chromosomes, the risk is 100 percent that any exposed offspring will have a genomic imbalance. With autosomal dominance, the risk is 50 percent that an offspring will be affected; with autosomal recessives, it is 25 percent. All these are very high risks when compared to contagious disease.

Third, both infectious and genetic diseases are subject to environmental variables. Malarial infections are more common in mosquito belts. The sickle-cell gene is more frequent in malarial regions.

Fourth, both infectious and inherited diseases are unequally distributed in different population and ethnic groups. The tuberculosis rate is extremely high among Eskimos and North American Indians, while Tay-Sachs disease is high among Ashkenazi Jews, and sickle-cell anemia is high among blacks.

Fifth, the morbidity and mortality rates vary among different infectious diseases and different genetic diseases. The common cold is a benign nonlethal disease, while rabies and the plague are almost always lethal. Webbed fingers and polydactyly are merely genetic nuisances. But Huntington's chorea and multiple polyposis are nearly always fatal.

Sixth, medicine has made great strides in the treatment of both infectious and genetic diseases. Pneumococcal pneumonia and gonorrhea can be effectively treated with antibiotics, while galactosemia can be treated simply by the elimination of milk sugar from the infant's diet.

And finally, some infectious and some genetic diseases can be prevented entirely. Polio and diphtheria have been nearly eradicated by vaccination, but continued surveillance and inoculation are necessary for each new generation of babies. Many genetic and chromosomal diseases are entirely preventable by continual surveillance of pregnancies by amniocentesis.

Now let us consider the social, ethical, and legal implications of the prevention and treatment of infectious disease and genetic disease. For simple, benign infectious diseases, few or no public health measures are necessary to protect society in general, and contacted individuals in particular. For malicious infectious diseases, strong public health measures are sometimes necessary, and these may infringe on individual freedoms and on invasion of privacy. For example, typhoid carriers are prevented from being public food handlers. Foreign travelers have been required by law to be vaccinated against smallpox. Persons with leprosy, infectious hepatitis, or tuberculosis may be required by law to be physically isolated from other members of society. Yet the courts have ruled that such a drastic measure as quarantine, which deprives the individual of his constitutional rights of freedom, is warranted. In these cases, public policy takes precedence over the individual. The rights of society are weighed against the rights of the infected individual. Two carriers of a lethal gene such as Tay-Sachs might be considered to be infectious to their offspring. One possible measure of prevention is genetic quarantine, such as sterilization, compulsory birth control, or abortion. Perhaps isolation of the gonads is not quite as great an infringement on individual freedom as isolation of the whole person. If such a public policy were adopted, it would favor protection of a certain segment of yet unborn children who may be dealt a certain death sentence if their parents are allowed to reproduce.

Many of the issues raised when we discuss genetic control are sticky ones, because they are emotional rather than social. They are not amenable to simple solutions. But I think we should consider controlling the spread of deleterious genes just as we have found it desirable to control the spread of pathogenic bacteria, viruses, and parasites. We need a basic premise on which we all agree when we discuss these issues. Perhaps we can adopt the goal that Bentley Glass has so eloquently stated: "The right of every child to be born physically and mentally healthy."

SUPPLEMENTAL READINGS

DHEW Publication No. (NIH) 75-370. 1975. What Are the Facts about Genetic Disease? National Institute of General Medical Sciences.

Fraser, F. C. 1974. Genetic Counseling. Am. J. Hum. Genet. 26:636-59. See also: Committee on Genetic Counseling. 1975. Genetic Counseling. Am. J. Hum. Genet. 27:240.

Glass, B. 1972. The Goals of Human Society. Bioscience 22:137.

Gleitman v. *Cosgrove*, 227 A.2d 689 (N.J., 1967).

Hsia, Y. E., and Silverberg, R. S. 1973. Genetic Counseling: How Does It Affect Procreative Decisions? Hosp. Prac. 8:52-61.

Jacobs v. *Theimer*, 18 Tex. Sup. Ct. J. 222 (1975).

McKusick, Victor A. 1975. Mendelian Inheritance in Man, 4th ed. Baltimore: Johns Hopkins Univ. Press.

GENETICS AND THE LAW

Riskin, Leonard L. 1975. Informed Consent: Looking for the Action. Univ. of Illinois Law Forum, pp. 580-611.

Roe v. *Wade,* 410 U.S. 113 (1973).

Schmickel, R. D. 1974. Genetic Counseling as a Form of Medical Counseling. Univ. Mich. Med. Center Bull. 40:38-43.

Shaw, Margery W. 1974. Invited editorial. Genetic Counseling. Science 184 (17 May):751.

Shaw, Margery W. 1975. Review of Published Studies of Genetic Counseling: A Critique. In Genetic Counseling Conference. New York: Raven Press.

Shaw, Margery W. 1976. Privacy and Confidentiality: Implications for Genetic Screening. Proceedings of 1st International Conference on Tay-Sachs Disease, Palm Springs, California, Nov. 30-Dec. 3, 1975.

Shaw, Margery W., and Damme, Catherine. 1975. Legal Status of the Fetus. In Genetics and the Law, pp. 3-18. Edited by A. Milunsky and G. Annas. New York: Plenum Press.

Stewart v. *Long Island College Hospital,* 296, N.Y.S. 2d 41 (1968).

Sutton, H. Eldon. 1975. An Introduction to Human Genetics, 2nd ed. New York: Holt, Rinehart and Winston.

DISCUSSION

Moraczewski: Obviously, one of the great concerns — one of the great ethical issues — is that of abortion. Many persons are concerned about it and each person has his own position regarding it, with which position Clarke College and others may or may not agree. We are trying to air the problems as clearly as possible and to hear the various arguments. It does not mean that we are espousing one or the other position on some of the ethical issues. I think it would be preferable not to focus on abortion *per se* as an ethical question. Recognizing that it is a problem, the Catholic Church is opposed to abortion, as I am. However, other ethical issues are more pertinent to this discussion. First I would like to ask Dr. Murray if he will respond to a few questions asked of him specifically.

Murray: I have a request to name "the most conservative theologian" who I said approved of induced abortion. I made that remark in relation to the Lesch-Nyhan syndrome, a condition affecting males where the individual is severely retarded and has an uncontrollable urge to cause pain and injure himself. The theologian I was referring to was Professor Paul Ramsey, a very staunch opponent of genetic abortion and so has been classified as conservative in that light. I would not say this of his other theological ideas. I was referring only to his position on genetic abortion. I don't believe he has stated his opinion publicly, so this has to be listed as a personal communication. He may have re-thought his position since I talked to him. The fact is that there is a situation (in Lesch-Nyhan syndrome) where the individual is in severe pain and is sincerely suffering. These are children who really want to be restrained. They have to have their arms tied and many of them must have their teeth removed because they even chew their own lips off. They are also often severely retarded. Ramsey thought this was a clear indication of conscious severe pain and suffering and, perhaps, this kind of life is better not lived. Again I hasten to say to you this was a personal communication and off-the-cuff discussion. Very often people re-think these informally stated positions, so he may not hold to this position. However, I think it important to make the point because if there is a valid argument about "wrongful" life, a life better not lived, perhaps Professor Ramsey would think this is such a life.

Another request: "You used the term predestination at the end of your presentation. Explain briefly what effect this would have on the future of genetic research in our society."

I used the term in reference to the fact that, as Dr. Carlson said, all of us carry gene mutations, and as Dr. Shaw said just now, several of them influence the effectiveness of our ability to reproduce ourselves. Many of you know the commercial where they say, "There's a new you coming every day!" Our cells are constantly being reproduced. But the reproduction of ourselves and of the DNA in the cells is imperfect. Mistakes are made—what we know in lower organisms to be somatic mutations. These are mutations that occur in the body cells and not in the germ cell and are not clearly established as occurring in humans, though a lot of the evidence points to their existence. And the fact that our reproduction is imperfect means that these imperfections will accumulate, and as a species, we will eventually deteriorate. Therefore, our demise is programmed into the genetic and cellular reproduction systems because the systems are imperfect. I do not mean predestination in the sense of what will happen to an individual in terms of his social or cultural future or what have you, or that the day of a person's death is already determined, but that the physical end of each individual is programmed in the imperfection of genetic and cellular reproduction. And this depends on which genes—whether it be genes affecting heart disease, genes affecting hypertension, as Dr. Shaw pointed out, or other conditions—are present.

Here is a request that is so complex that I can really only scratch the surface. It relates to a 1975 Nobel Conference at which Dr. William Shockley was present and apparently made a statement that black people are genetically inferior to white people. To respond properly to this would require another symposium all by itself. Briefly stated, one of Dr. Shockley's contentions is that there is a reduced social adaptability of blacks, not only blacks but also Spanish-speaking individuals, based on a social adaptability index he devised, using as a basis the listing of various ethnic groups in American Men of Science, their presence in the National Academy of Science, and the number of persons of a given racial group who won Nobel prizes, in contrast to the frequency of these groups in prisons, their illegitimacy rate, frequency of drug addiction, etc., all of which he concludes, as some early geneticists did, are mostly genetically determined. The social adaptability index was predestined to predict what he found, namely, that the blacks and Spanish-speaking individuals are inferior because they have a lower social adaptability than do Caucasians; on the other hand, he does not usually point out in his public addresses that his index should also predict that Orientals and Jews are superior to the average European white Anglo-Saxon Protestant. So he uses the index as he wishes. There are other predictions he makes based on IQ scores. These are very complex issues and cannot be discussed at this time.

Carlson: I'd like to ask Dr. Shaw about the policies that the state or the federal government may adopt on genetic laws. Most geneticists fear bad laws. They prefer no legislation to poor legislation. But most legislators are not biologists, do not have a medical background, and do not understand genetics. I wonder what can be done to raise the awareness of persons who make laws, because at present they aren't familiar with the field.

Shaw: I had a draft of a law that a congressman in another state (not Iowa) sent to me for comments and suggestions in which he was recommending that all persons applying for marriage licenses be seen by a genetic counselor, and if the risks of having defective children were greater than 50 percent the marriage license be denied. This suggested the ignorance of Mendelian laws, as well as the problems of implementing such a public policy. I think it is a terrible problem but I don't know that it's any worse for geneticists than for economists, or environmentalists, or fuel experts, or whatever. I think we are going to have to live with congressmen who don't know about everything and hope their staff, in drafting legislation, will go to individuals who can give them some sound advice. Don't interpret the statement to mean that I am drafting a lot of genetic legislation right now, but, if it were to come about, I don't see any other mechanism. I don't think we can educate the senators and the congressmen to understand genetics in a two-hour crash course.

Walters: I think that what the state of Maryland is doing right now is a very interesting experiment. Maryland has set up a State Commission on Hereditary Disorders, rather than rushing into legislation. This commission has been at work for about four years and includes geneticists, physicians, and politicians; the members have worked through individual diseases like phenylketonuria (PKU), trying to decide whether it would be justifiable to pass a law requiring mandatory PKU screening in Maryland. In the case of PKU the commissioners decided that mandatory screening is justified, but they are proceeding on a case-by-case basis, disease by disease, trying to decide what the pros and cons are and whether it would be wise on balance to pass a law with respect to each specific disease. I think the time allowed for study and reflection in this process is very important. The commission's deliberate pace contrasts sharply with the haste that sometimes accompanies genetic legislation.

Moraczewski: Dr. Shaw, you made the remark that traditionally primitive societies did not recognize the fetus as a person. I wonder what evidence you have to support that statement.

Shaw: Certainly it is not a statement that is all black and white. From a review of some of the societies in the South Sea Islands, from some of the primitive societies in Africa, and from a review of the literature on some of their attitudes toward contraception, abortion, and infanticide, it is seen that many of these societies had a problem of limitation of food resources. Also societies that did a lot of migrating had different attitudes from those settled, civilized or precivilized societies who were in an agricultural environment.

Moraczewski: One of the difficulties is that the notion of person, of personhood, is relatively recent in terms of the human race. Even the philosophers such as Plato and Aristotle did not clearly distinguish between the notion of person and human nature. And the notion of person seems to develop more out of Christian theology, when the problems of Christology and the Trinity became a burning issue. One needed to distinguish between person and nature, human person and human nature. And that, as far as I know, was the origin of the clear distinction between the two. So it is not surprising that in many cultures the notion of person is not clearly defined today and this leads to some of the remarks you made about legal rights and legal personhood in the Supreme Court's decision on abortion. It is very important in our discussion to distinguish between a legal person and a natural person, or an ontological person. The Court, as such, certainly is competent to declare whether a person is legal — legal, for example, in terms of his right to vote. An individual is a legal person now at the age of 18 but the question is: Is the Court competent to define legal personhood when it comes to rights that are antecedent to the Supreme Court? For example, our founding documents speak about the inalienable rights of every individual—the right to life, liberty, and the pursuit of happiness—and you wonder whether the Supreme Court would have the power to remove these rights from an individual. And so I say the area needing constant clarification is the distinction between a legal person and a natural or an ontological person. That's where a lot of discussions fall into the morass of confusion, due to a failure to make that distinction. The consequences, of course, are enormous while we decide about the question of the person of the fetus—a difficult, complex situation. I think the conflict sometimes arises between the rights of two persons, and more energy should be devoted to how you resolve a very difficult conflict situation. How do you resolve that justly? That's better than trying to solve the problem by eliminating one person from the discussion; that might be a more fruitful direction for the discussion.

Shaw: I certainly agree with Father Albert that much of our dilemma in these discussions is that we do not start by saying, "I am now defining a biological person or a human zygote," or "I am defining an ethical being, a religious being," or "I am defining a legal being." I might point out that in the law there are certainly changes that do occur. Two very pronounced changes in the recent evolution of constitutional interpretation have been in the rights of minors compared to adults and in the rights of illegitimate children—their legal rights compared to those of legitimate children. So I think there are two trends in the law in which changes in society have promoted changes in legal thinking. The whole question of the being of the individual *in utero* and the input from society concerning society's opinions about this will be reflected in legal opinions.

Murray: Dr. Shaw, you pointed out similarities between genetic and infectious diseases. Would you like to point out at least one or more differences between the two because I think the differences are highly significant, maybe in some respects even more significant than the similarities?

Shaw: The differences that are most frequently commented upon are a time difference and the number of individuals involved in the risk. In terms of the time span, if you have a sudden crisis of an infectious disease right now that may spread throughout the community and then out into the world, it would require on the part of public health authorities an immediate reaction rather than a reflective or a considered reaction. It might require very stringent action in getting the disease under control. The other difference that is frequently commented upon is that infectious diseases may expose a very large number of people rather than exposing only one individual as in the case of genetic disease. Again, this is only temporal because in the case of venereal disease for which we have many laws regulating its spread, it is a one-to-one contact.

Murray: There is one other difference that is important but may not be a difference in the development of future technology, namely, the possibility that in the case of infectious diseases the offending agent can be eliminated from the body or at least neutralized and have no effect. When one has a genetic defect, a mutant gene, that cannot be eliminated without eliminating the person. And in the words of that famous cartoon character, Pogo, when we face the enemy, in this case, it is US. In trying to fight genetic disease, fight Huntington's chorea, fight sickle-cell anemia, you're combatting in a certain sense other human beings or potential human beings. This makes it very difficult to advocate contagion, isolation, or what have you as measures for controlling the genetic disease. Those are the differences that make the analogy between the two concepts very tenuous and that make it somewhat dangerous when you apply preventive measures.

Walters: In defense of Dr. Shaw on that point, wouldn't there be prevention if the counseling and decision were made prior to conception? In other words, you don't eliminate any person or potential person if the couple decides not to bear children unless there is a right of the unconceived to be conceived!

Murray: Let's take an example: If a room was filled with people of Italian ancestry, there would be a certain frequency that one person would have one of the thallesemia disorders and somebody else would have one of the other blood disorders common to the Caucasian, Italian, or Greek ancestry. I don't know how many, knowing that, would still say they should not have been conceived. In other words, maybe there is such a right, highly theoretical though it may be, of the unconceived to be conceived even though that life be short. It is that view that has made Arlo Guthrie decide not only that he would have children but feel that his children should have children with Huntington's disease, because even though such patients live only

30-40 years and their demise is very unpleasant, nevertheless, they do live 30-40 years with the opportunity to contribute to society and culture although the genetic defect itself is not a very pleasant one. This was one of the things I was alluding to when I was talking about making a judgment about the value of human existence. Are we to say that unless someone is guaranteed 65-70 years of existence that person does not have a right to be conceived on this earth? I think some people would say absolutely not. Even if they are guaranteed only 5 years of existence, they have the right to be conceived. And this is one of the things I am most concerned about when talking about isolation from genetic disease contagion.

Moraczewski: This points out something very necessary that was brought out briefly about social values — social goals that we should have. We need to be articulate because as Dr. Murray pointed out, what do we really want as a society? Is it better for a person to have a long life but with disease, to have a short life with no disease at all, or not to have been born? I was amazed in this connection to read a comment by Dr. Cecil Jacobson from George Washington University who was quoted in an article as saying that if it were possible to diagnose in the fetal state that an individual would have cancer by the age of 40 or 50, he would abort that fetus. That sounds rather incredible but is stated as a goal. If you begin to think in those terms, you can press that rather far and there would be no people alive; for, as Dr. Carlson said, "No one dies of old age."

Carlson: Parents should make the decision as to whether or not a child is to be born. Their decision is going to be based on their expectations and involvement with the child. If they had foreknowledge that the child to be born was going to devastate their marriage and exclude almost every other activity except the care of a chronically ill child, they might decide not to have children. So I don't believe there is an absolute right for a child to be conceived. If this is the case, then the decision to have a child requires some sort of counseling or knowledge that has to come from some informed source. If there are genetic defects, that information should be made available. Every pair of parents will differ in the way they arrive at a decision. But I feel that's fine in a democracy — that we should provide as much information as possible. If they make what geneticists or physicians consider bad choices, that is still acceptable because they are free individuals to make that choice. But at the same time, I feel as a geneticist that I have an obligation to make them as informed as possible about the things that can go wrong because the most mischief that can be done, in my mind, is not letting people know the consequences of their "right to conceive."

Walters: Dr. Shaw, do you think that the pain involved in learning about one's genetic disorders, or one's carrying a certain recessive genetic trait, is qualitatively different from the pain of learning anything medically adverse about ourselves? For example, I'm thinking of learning that one has cancer and of the pain and the trauma that this new knowledge might cause. Is there something qualitatively different about a genetic disorder and genetic information?

Shaw: I think there are differences and similarities. I don't know if I can summarize my own position. The trauma of finding out that your child has leukemia and is not responding to treatment and the life span is 6 months or under is a very psychologically painful experience and parents need a great deal of support to cope. A sudden automobile accident that wipes out a loved member of the family is another. One thing is, we might have some built-in coping mechanism, derived from past experiences in having seen other people going through these traumas, that might prepare us a little bit more than the sudden awareness that it is one's own genes—our own defects—that are causing the defective child. And that adds a dimension of guilt that, say, getting a disease or being killed in an auto accident doesn't have. I've often thought that people in a medical setting respond more emotionally to disease of the brain and the gonads and possibly the heart than to any others; that, if you have a kidney disease or a muscle disease, it somehow is not quite as stigmatizing as having a brain disease or being told you have a defect in your ovaries or testes that is going to create pain and suffering in other individuals. So there are some similarities certainly in the coping process, and how to handle it once you've been told is similar, but there may be something qualitatively different.

Murray: I think it has become apparent, particularly in the mass scare surrounding the sickle-cell question, that although there are similarities in that people are concerned about "what did I do to cause this," there are differences because we do not ask for our genes, they are given to us, so to speak. They are something over which we have no control. Not only that, if the genes have been passed down for many generations, one of the cultural ideas we have about Providence preordaining punishing subsequent generations for their sins and so forth becomes a very important thought in the minds of many people. In our genetics clinic we have spent a lot of time trying to convince people who have had one or more children with sickle-cell anemia that they are not being punished for something they did. Sometimes it is impossible to help them, to rid them of the idea that they have committed some sin for which they are being punished and for which their child is being punished. This is very hard for some guilt-ridden people to accept in their coping process. There is a difference between the small child who gets leukemia and the small child who gets a disease for which a husband and a wife passed on the gene. These feelings are also culturally strong. They are ingrained so much that, in the study of Dr. John Fletcher concerned with couples who were waiting to hear the results of amniocentesis for Down's syndrome, the parents who got the normal results were, of course, very happy, but the mothers in the one or two cases found to have another child with Down's syndrome developed strong feelings of what he called cosmic guilt. Even though the geneticist might say, "Well, those are the chances, let the chips fall where they may, it turned out bad for you," the mothers in those cases had strong feelings that it wasn't just chance, that somebody—God, some cosmic force, or what have you—made that happen to me. It is this kind of thing that is very hard to cope with and is especially hard in genetic disorders because of the cultural milieu we come from.

Moraczewski: That becomes a theological question. Why is God doing this to me? It is a question that permeated the Bible. We find Christ responding to it when He was asked, "Did this man sin or did his parents sin?"—referring to a particular individual who was brought to Him. He responded that neither sinned. This response needs to be brought across emphatically. Perhaps the chaplain or the religious counselor can really contribute to a situation like this.

4/

Genetics, Reproductive Biology, and Bioethics

LEROY WALTERS, Ph.D.

IN 1932 Aldous Huxley published a futuristic novel, which included the following passage:

> "I shall begin at the beginning," said the [Director of Hatcheries and Conditioning]. . . . "These," he waved his hand, "are the incubators." And opening an insulated door he showed [the students] racks upon racks of numbered test-tubes. "The week's supply of ova. Kept," he explained, "at blood heat; whereas the male gametes," and here he opened another door, "they have to be kept at thirty-five instead of thirty-seven. Full blood heat sterilizes. . . ."
>
> Still leaning against the incubators he gave them . . . a brief description of the modern fertilizing process; spoke first, of course, of its surgical introduction—"the operation undergone voluntarily for the good of Society, not to mention the fact that it carries a bonus amounting to six months' salary"; [the Director] continued with [an] account of the technique for preserving the excised ovary alive and actively developing; . . . referred to the liquor in which the detached and ripened eggs were kept; and, leading [the students] to the work tables, actually showed them how this liquor was drawn

---

LEROY WALTERS received his graduate education in ethics at the University of Heidelberg, the Free University of Berlin, and Yale University (Ph.D., 1971). Since 1971, Professor Walters has been director of the Center for Bioethics at the Kennedy Institute, Georgetown University. He has served as a consultant to the National Commission for the Protection of Human Subjects and is a member of the Recombinant DNA Molecule Program Advisory Committee at the National Institutes of Health. Professor Walters is editor of the annual *Bibliography of Bioethics* (Detroit: Gale Research Company, 1975-77) and coeditor of a recent anthology entitled *Contemporary Issues in Bioethics* (Encino, California: Dickenson Publishing Company, 1978).

off from the test-tubes; how it was let out drop by drop onto the specially warmed slides of the microscopes; how the eggs which it contained were inspected for abnormalities, counted, and transferred to a porous receptacle; how . . . this receptacle was immersed in a warm bouillon containing free-swimming spermatozoa—at a minimum concentration of one hundred thousand per cubic centimetre, he insisted; and how after ten minutes, the container was lifted out of the liquor and its contents re-examined; how, if any of the eggs remained unfertilized, it was again immersed, and, if necessary, yet again; how the fertilized ova went back to the incubators, where the Alphas and Betas remained until definitely bottled; while the Gammas, Deltas, and Epsilons were brought out again, after only thirty-six hours, to undergo Bokanovsky's Process.[1]

In September 1973 there appeared in the British journal *The Lancet* a brief article entitled "Transfer of a Human Zygote." The authors reported that a human ovum had been fertilized in a test tube, then transferred to a special medium for further development. After 49 hours the zygote had divided into three cells. Eighteen hours later the zygote had reached the eight-cell stage. At the 74-hour mark the early embryo was transferred into the uterus of the ovum donor through a small polyethylene tube. Biochemical tests during the next several days seemed to indicate that the embryo had implanted, but on the eighth day the embryo spontaneously "aborted," and the experiment failed.[2]

Within the field of bioethics a great deal of controversy has surrounded several reproductive technologies. Perhaps no technology has received more attention than test-tube fertilization, the union of sperm and ovum *in vitro*.

Two polar positions on test-tube fertilization have emerged. On the one hand, Professor Joseph Fletcher welcomes new and artificial modes of reproduction as a much-needed alternative to what he calls the traditional coital-gestational method. Indeed, Fletcher argues that only by employing such new methods will we be able to "end reproductive roulette" and begin to reduce our overwhelming load of genetic defects.[3]

On the other hand, Dr. Leon Kass and Professor Paul Ramsey have strongly condemned the use of test-tube fertilization as a reproductive technique. Dr. Kass has noted that the technique is experimental and could easily result in deformed offspring. He therefore raises the question whether test-tube fertilization does not constitute unethical experimentation on the unborn.[4] Professor Ramsey views fertilization of human ova in test tubes as a giant step toward the laboratory control of human reproduction. In the words of Ramsey,

> We shall have to assess *in vitro* fertilization as a long step toward Hatcheries, [that is,] extracorporeal gestation, and the introduction of unlimited genetic changes into human germinal material while it is cultured by the Conditioners and Predestinators of the future.[5]

Is it possible to steer a middle way between these polar positions and to build a moral argument for the use of test-tube fertilization under certain curcumstances? I believe that such a middle course is possible and would like to ask you to accompany me on a brief thought-experiment. In this thought-experiment we shall make two assumptions. The first is that the desire of married couples to have children of their own is a reasonable desire. The second assumption is that human life begins at the time of fertilization and that therefore the developing embryo deserves to be protected.

Let us begin our thought-experiment with Case 1. A couple desires to have children but discovers that the husband's sperm count is too low to allow fertilization to take place following sexual intercourse. A physician therefore suggests that specimens of semen be collected from the husband over a period of several weeks or months. The couple agrees. The specimens are collected, preserved by cryobanking, then combined and concentrated. The resulting concentrate is used by the physician to inseminate the wife, who conceives and bears a child.

I suspect that most of us would consider that the couple was morally justified in resorting to artificial insemination in this case. Procreation is separated from sexual intercourse, it is true, but only when the couple has determined that sexual intercourse does not lead to procreation in any case. The physician functions as a kind of early midwife, helping the couple with the beginnings of the pregnancy rather than with the delivery of a fully developed fetus.

Let us proceed, then, to consider Case 2. In this instance the tables are turned. The wife is infertile because of blocked Fallopian tubes. Since female sex cells are less accessible than the male's, a minor surgical procedure is required to remove several ova. The attending physician brings one of the ova into contact with the husband's sperm in a test tube and transfers the resulting embryo to the uterus of the woman, where the embryo implants and grows to maturity.

This case may give us pause because of the test tube and because of the more active role played by the physician. But Case 2 closely parallels Case 1. In both cases sex cells are moved to the site of fertilization. It is merely the mechanics of reproductive biology that require the sex cells of the two partners to be united outside the woman's body. Indeed, one might justly be accused of discriminating against women if one approved artificial insemination in Case 1 and disapproved test-tube fertilization in Case 2.[6]

Case 3 takes us one step further. Let us suppose that both a husband and wife can produce normal sex cells but that the wife suffers from a uterine disease or insufficiency making it impossible for her to carry a fetus to term. If a safe, proven artificial womb were available, would the couple be justified in requesting that test-tube fertilization be performed using their sex cells and that the resulting embryo be transferred to the artificial womb — whence it would be delivered after the normal 240 days of gestation? If one were willing to approve the couples' actions in Cases 1 and 2, it would be difficult to find grounds for disapproving the extracorporeal gestation of Case 3.[7]

What, then, is the point of this thought-experiment? It is not to advance a slippery-slope argument. It is, rather, to indicate that *new reproductive technologies are not, in and of themselves, to be feared*. Many of these technologies, including test-tube fertilization, may be of great long-term benefit to the human race. In my view, these new technologies should be developed and should be employed to alleviate human need.

The real danger of futuristic reproductive technologies lies not in the technologies themselves but *in the social and political uses to which they may be put*. Thus the same techniques that could be of assistance to future childless couples also play a central role in the Hatcheries of Huxley's *Brave New World*. Everything depends on our wisdom and the wisdom of future generations in properly harnessing the potentialities of the new technologies.

Now I would like to shift our focus to the present, to two genetic techniques already widely used. The first is postnatal genetic testing through the analysis of blood, urine, or tissue samples; the second is prenatal diagnosis by means of amniocentesis.

Every year postnatal genetic testing grows more sophisticated, and the list of detectable disease and recessive traits expands. Mass screening for genetic disease began in the early 1960s and Massachusetts was the first state to establish a mandatory program. In 1963 Massachusetts began compulsory testing of newborns for phenylketonuria (PKU). Since 1963 most other states have followed suit. Some states, including Massachusetts and New York, now require the testing of newborns for a whole series of genetic diseases.[8]

The general rationale for these mandatory programs of screening for genetic disease is that afflicted newborns can be detected early and can thus be given prompt treatment for their metabolic defects. I believe that most of us would find this rationale morally acceptable, since newborns would probably give ready consent to such testing if they were able to understand the potential benefits to them of the screening programs.

A qualitatively new stage in postnatal genetic screening was reached in the late 1960s, when programs were instituted that sought to detect heterozygous *carriers* of specific recessive traits rather than homozygous *sufferers* from particular genetic diseases. The two major traits screened were those for Tay-Sachs disease and sickle-cell anemia. In the case of Tay-Sachs disease all carrier-screening programs were voluntary; however, several states passed laws requiring the establishment of statewide programs to screen for carriers of sickle-cell trait.[9] Here the rationale for mandatory screening programs was less clear. It was probably not to restore the carriers of sickle-cell trait to health, since carriers generally do not suffer ill effects from the recessive gene they carry.[10] Insofar as there was a clear rationale for the mandatory screening of sickle-cell trait, it seems to have been to influence marital and reproductive decisions by carriers.

No new technologies would be required to expand mandatory postnatal screening to total coverage of a nation's entire population. Joseph Fletcher has described how a comprehensive national program might work:

> A socially conscientious system would be a national registry; blood and skin tests done routinely at birth and fed into a computer-scanner would pick up all anomalies, and they would be printed out on data cards and filed; then when marriage licenses are applied for, the cards would be read in comparison machines to find incompatibilities and homozygous conditions.
>
> The objection is, predictably, that it would "violate" a "right"—the right to privacy. It is even said, in a brazen attack on reason itself, that we have a "right to *not* know." Which is more important, the alleged "privacy" or the good of the couple as well as of their progeny and society? (The couple could unite anyway, of course, but on the condition Denmark makes—that sterilization is done for one or both of them. And they could even still have children by medical and donor assistance, bypassing their own faulty fertility.)[11]

GENETICS, REPRODUCTIVE BIOLOGY, AND BIOETHICS

Parallel developments in practice and theory can be noted in the field of prenatal diagnosis by means of amniocentesis.[12] The technique of amniocentesis is presently employed to diagnose fetal chromosome abnormalities and approximately 50 different types of fetal genetic disorders.[13] On the horizon, and already in use experimentally, are two additional diagnostic techniques—the use of the fetoscope to visualize the fetus *in utero* and to assist in securing fetal blood samples from the placenta.[14,15]

At present there are no mandatory programs of prenatal diagnosis in the United States. Women, particularly women at high risk for producing offspring with genetic disorders, enter into genetic counseling on a voluntary basis and receive amniocentesis only at their request. More comprehensive programs of prenatal screening by means of amniocentesis have been proposed, however. For example, a 1973 article in the *Lancet* outlined a "Screening Programme for Prevention of Down's Syndrome."[16] The article proposes that amniocentesis be offered to every pregnant woman and that induced abortion be provided on a voluntary basis to every woman carrying an affected fetus. In support of this proposal the authors note that "the total prevention of Down's syndrome could reduce the prevalence of severe mental retardation by 30%," whereas "the total prevention of PKU by current means could reduce the prevalence of severe mental retardation by [only] about 1%."[17]

To my knowledge, mandatory genetic screening programs requiring either amniocentesis or eugenic abortion have not been proposed. Even Joseph Fletcher stops short of explicitly espousing such programs, although some of his general comments on compulsory screening might conceivably be applied to programs of mandatory amniocentesis or mandatory abortion. Consider, for example, the following passage from *The Ethics of Genetic Control*:

> Our moral obligation to undergo voluntary screening, if it is indicated, is too obvious to underline. The squeeze here, ethically, is that the social good often requires *mass* screening. When it is voluntary, it is "nicer," as we see in the popular acceptance of tests for cervical cancer. But let it be compulsory if need be, for the common good—[Garrett] Hardin's "mutual coercion mutually agreed upon."[18]

Between now and the Tricentennial it seems likely that furious battles will be fought over the issues of mandatory postnatal and prenatal screening. Indeed, one can almost sense that the rhetorical weapons for this battle are already being chosen. Proponents of compulsory screening programs will employ public-health analogies, noting that just as the horizontal transmission of infectious disease must be contained, so must the vertical transmission of genetic disease to future generations be prevented. They will point to existing laws requiring premarital blood testing for venereal disease or immunization against smallpox or rubella. State laws prohibiting marriage for the retarded or permitting compulsory eugenic sterilization of the mentally handicapped will also be cited. Finally, the long-term financial cost of not instituting mandatory screening programs will be calculated, particularly if the society as a whole is asked to foot the bill.[19]

Opponents of mandatory screening, for their part, will be armed with a battery of counterarguments. Compulsory screening programs that would single out particular ethnic groups for the sake of cost-effectiveness will be attacked as violations of the equal protection guaranteed by the Constitution.[20] It will be argued that the cost-benefit analysis of the mandatory-program advocates is too narrow, that it focuses entirely on economic costs and overlooks the heavy toll of human suffering that compulsory programs inevitably entail.[21] Finally, opponents of mandatory screening programs will speak of fundamental human rights that ought not be violated, no matter how appealing the goal. They will appeal to the constitutional right of privacy and will assert that it implies a right to marry and procreate and a right to exercise control over one's own body.[22] It would be a strange quirk of irony, indeed, if one day the *Roe* and *Doe* decisions of the Supreme Court were invoked to defend a woman's right to privacy against state-imposed programs of mandatory amniocentesis or abortion.

We have briefly surveyed two biomedical technologies—test-tube fertilization and two subtypes of genetic testing. I shall conclude by drawing several comparisons between the two technologies and by attempting to formulate a few general conclusions.

Test-tube fertilization is still a future prospect.[23] Genetic testing, both postnatal and prenatal, is already a widely used technique. Test-tube fertilization has been widely discussed, even sensationalized, in both science fiction and ethical analyses. The story of genetic testing has been less publicized, perhaps because the technique has been developed incrementally. I have tried to argue that the use of test-tube fertilization could be morally justified under certain circumstances, and we have seen that genetic testing, when linked with programs of mandatory screening, could seriously undermine cherished constitutional freedoms.

Here we come back to a point noted earlier. The nature of the technology itself is less important than the social and political uses to which the technology is put. The techniques of genetic testing that make possible an ethic of genetic responsibility also could be used literally to determine which types of genes will be tolerated in the human gene pool.

The need of the hour is for the largest possible group of informed citizens who will be alert to new technological possibilities and who will — in a rational, balanced way — seek to assess the potential social impact of those possibilities. Only with the aid of such careful technology assessments will we be able to set social priorities that both promote the human good and enhance human freedom. And only by setting our social priorities properly will we be able to hand on a worthy heritage to the Tricentennial People.

NOTES

1. Aldous Huxley. 1946. Brave New World, pp. 3-4. New York: Harper & Brothers.
2. D. De Kretzer et al. 1973. Transfer of a Human Zygote. Lancet 2(7831):728-29.
3. Joseph Fletcher. 1974. The Ethics of Genetic Control: Ending Reproductive Roulette, pp. 64-70, 165-66. Garden City, N.Y.: Anchor Books.
4. Leon R. Kass. 1971. Babies by Means of In Vitro Fertilization: Unethical Experiments on the Unborn? New Eng. J. Med. 285(21):1174-79.
5. Paul Ramsey. 1972. Shall We "Reproduce"? J. Am. Med. Assoc. 220(11):1481.
6. I am indebted to my colleague Dr. André E. Hellegers for this perspective on test-tube fertilization. Dr. Hellegers terms the differential evaluation of artificial insemination and test-tube fertilization "male chauvinist piggery."
7. I shall not take up the more difficult question whether the techniques described in Cases 1, 2, and 3 should be employed for the sake of safety or convenience even in cases in which no infertility problem exists. Artificial insemination using donor sperm, as well as embryo transfer to a host mother, would raise complex issues of fourth-party involvement. For an account of thirteen "rungs" in the ladder of test-tube reproduction, see Gerald Leach, The Biocrats, pp. 69-98, New York: McGraw-Hill, 1970.
8. Tabitha M. Powledge. 1974. Genetic Screening as a Political and Social Development. In Ethical, Social and Legal Dimensions of Screening for Human Genetic Disease, pp. 30-32. Edited by Daniel Bergsma. New York: Stratton Intercontinental Medical Book Corp.
9. Ibid., pp. 33-34.
10. Ibid., pp. 36-37. The data on this point are somewhat ambiguous but in general point in the direction noted.
11. Fletcher, Ethics of Genetic Control, pp. 182-83.
12. Prenatal diagnosis of some fetal conditions can also be carried out by means of ultrasonography or by analysis of maternal blood or urine samples. However, amniocentesis is the most widely employed technique.
13. Barbara K. Burton, Albert B. Gerbie, and Henry L. Nadler. 1974. Present Status of Intrauterine Diagnosis of Genetic Defects. Am. J. Obstet. Gynecol. 118(5):718-46.
14. J. E. Patrick, J. B. Perry, and R. A. H. Kinch. 1974. Fetoscopy and Fetal Blood Sampling: a Percutaneous Approach. Am. J. Obstet. Gynecol. 119(4):539-42.

15. John C. Hobbins and Maurice J. Mahoney. 1974. In Utero Diagnosis of Hemoglobinopathies: Technic for Obtaining Fetal Blood. New Eng. J. Med. 290(19):1065-67.
16. Zena Stein, Mervyn Susser, and Andrea V. Guterman. 1973. Screening Programme for Prevention of Down's Syndrome. Lancet 1(7798):305-10.
17. Ibid., p. 308.
18. Fletcher, Ethics of Genetic Control, p. 182.
19. Powledge, Genetic Screening, pp. 38-39; cf. Harold P. Green and Alexander M. Capron. 1974. Issues of Law and Public Policy in Genetic Screening. In Bergsma (ed.). Ethical, Social and Legal Dimensions of Screening, pp. 68-70.
20. Green and Capron, Issues of Law and Public Policy, pp. 76-77.
21. Daniel Callahan. 1973. The Meaning and Significance of Genetic Disease: Philosophical Perspectives. In Ethical Issues in Human Genetics: Genetic Counseling and the Use of Genetic Knowledge, pp. 86, 89. Edited by Bruce Hilton et al. New York: Plenum Press.
22. Green and Capron, Issues of Law and Public Policy, pp. 71-76.
23. There have been no documented instances of successful human test-tube fertilization leading to the birth of a child.

DISCUSSION

Shaw: Dr. Walters, will you elaborate on or perhaps clarify the positions of Dr. Kass or Dr. Paul Ramsey on taking the chance of a deformed baby by *in vitro* fertilization? Did they differentiate a moral difference between taking a chance by AID [artificial insemination by donor] or taking a chance of having a deformed baby by natural sexual reproduction?

Walters: In both of them there is a very strong commitment to natural sexual reproduction. I think that they view the risk of having a deformed offspring through the traditional mode of sexual reproduction as unavoidable.

Moraczewski: Is it possible for genetic mutations to be not defects but an advantage to an individual or a race? Presumably, in an evolutionary interpretation this has happened. Currently, when we speak about genetic mutations, these genetic changes are always considered to be of negative value. Could some be of positive value?

Carlson: Most mutations that arise spontaneously have the same adaptive value as occurs when you are typing a paper and introduce a typographical error. The chances that this improves the sentence, the paragraph, or the whole term paper or that it changes the grade are close to nil. This is approximately what happens when a gene undergoes mutation because the chances that any random disturbance of the sequence that has been built up out of selection over

GENETICS, REPRODUCTIVE BIOLOGY, AND BIOETHICS

thousands of generations would be altered in a way that improves it is unlikely. The chances are overwhelming that the function of the protein made by that gene will be defective.

Moraczewski: There still remains, of course, the terminus of evolutionary development. How do you explain that?

Carlson: Again natural selection explains that. Occasionally, these lucky things do happen. There are times when I do type a happier word by typographical error than I had originally thought to do, but rare is the day when that happens! The same is true of gene mutations. Occasionally, an improvement is made and it is these rare events that permit the evolution of improvements and complexity. But these rare beneficial mutations require a sifter to get rid of the far larger number of defective genes in the population. In the past, infectious diseases played a major role in eliminating defective genes in the population. But if we don't have natural selection in man any more, then we must face the reality that the overwhelming amount of gene mutations is to our detriment.

Shaw: One question asked of me is: What is Huntington's chorea? It is an autosomal dominant genetic disease in which the individual is born perfectly normal and healthy and grows up to an adult life and sometimes, usually around the age of 25-30-40, begins to develop symptoms of degeneration of the nervous system. This is manifested by mental changes, loss of memory, some psychiatric disturbances, and primarily by involuntary muscle movement called choreiform movements that affect the arms and legs and all the skeletal muscles, including the facial muscles, tongue, and swallowing muscles. This is continuously built up; the disease gets progressively worse; the individual usually lives in a debilitated condition for 5-10 years, confined to a wheelchair, loses mental capacity, and eventually dies, often of pneumonia, because of the inability to swallow food without choking on it, or gets an infection and dies. The disease is not diagnosed by amniocentesis. It cannot be diagnosed in the child or young adult before the onset of the symptoms, and yet the probability is 50-50 that the offspring of any affected individual, either male or female, will have the gene and if he/she lives into the 40s will get the disease.

Now I have three questions relating to law. The first is: "In the near future do you see any possibility of certain genetic procedures becoming so involved as to need legal restrictions?" I don't know what the question means—to become involved—does it mean to become complicated technologically or to become involved in the privacy of the individual? As a general answer I certainly agree with Dr. Carlson that no laws at all are better than bad laws and a legal restriction of any kind — whether it be genetic, reproductive, or an infringement on human personal freedom and on our democratic process — should be very seriously considered before such things come to pass.

Second, the comment was made that my reference to the legal status of the fetus was derived from English and American law, and what about the Roman law and the Napoleonic code, which are more widespread in the world? Are the principles of the legal status of the fetus the same? Roman law or civil law is certainly more widespread than English common law. The laws in our country with the exception of the state of Louisiana are derived from the British common law system and the Louisiana state laws are derived from the French civil system. I am not familiar with the statements made about the fetus in the Roman civil law. If someone is expert in classics or in history or in embryology of the middle centuries, I would really appreciate help on that question.

This next question is concerned with the fact that in the early days the fetus was considered only as property rights so those fetuses who were from poor families had no rights at all. I guess that one could say that since they were recognized only as property rights they had no biological rights at that time recognized in the court. Are they victimized by the law? Certainly, fetuses, illegitimate children, and minors have been victimized by our laws for a long time.

Moraczewski: One question is addressed to all members of the panel. Are you or your institutions financed by the March of Dimes, and if so, in what way?

Carlson: No, I don't know of any of my colleagues at the State University of New York at Stony Brook who are financed by the March of Dimes. The only contact I have had with the March of Dimes is its educational program. It does have a list of free literature that is provided to anyone who writes in. Articles are reprinted from professional journals and I have taken advantage of this free educational service to obtain information on different disorders so I can familiarize myself with them.

Shaw: At our institution there is no support for the Medical Genetics Center. However, the pediatrics department has about a $2,000-a-year grant from the March of Dimes to train nurses and social workers to take family histories.

Moraczewski: I do not get any support from the National Foundation or March of Dimes.

Murray: Our genetics division is not supported in any way, but the department of obstetrics at Howard Medical School has a 2- or 3-year grant from the March of Dimes to study the factors surrounding the birth of high-risk infants and those that have birth defects either from environmental causes or from genetic causes. This department is also examining the medical and prenatal histories of mothers whose children are born prematurely or with deformities or with other kinds of birth defects.

Walters: Our institute receives no support from the National Foundation. The National Foundation did establish a Bioethics Advisory Commission in 1975 and appointed about eight people in medicine, law, and ethics to be on that advisory committee. I am a member of that committee.

Moraczewski: Other general questions on the educational level are: What is required for a master's degree in genetic counseling? What kind of screening would be done prior to admission to such a course? Would this involve genetic screening?

Murray: There are three programs, and each program is different. They have somewhat similar principles in that they are affiliated at centers where genetic research and counseling are being done and where there are genetic clinics. But, as far as I know, they have no stringent prerequisites other than that the person have a bachelor's degree and have a reasonable academic background with the aptitude and desire to do the work. Most of the students participating have been women; a large percentage are women who got their degrees some years ago, had a family, and are now interested in going back to work. And many of them are people who have been in the behavioral sciences—have degrees in social work or psychology. Most of them have been able to find work, although only a minority have positions where they are doing what they are trained to do right now, that is, genetic counseling. That's because there is a shortage of funding for these positions around the country. This situation, I believe, is slowly changing.

Moraczewski: There are several questions in regard to cloning that ask primarily for a definition or description of cloning.

Carlson: Cloning can be done in several ways. One of the simplest ways is to tap some cells from an already existing organism, for example, a carrot. You take the carrot root and cut off a coin-shaped disk and then cut it up into small slivers. Some enzymes are then added to digest away the cell's adhesive properties and the individual cells can float free. A single cell can then be cultured in a test tube to produce a new carrot. This implies that every cell, even though it can be differentiated into different tissues in the carrot, is capable of producing an identical twin of itself. In animals this can be done in early embryos. You can take a 16-cell stage and slice it up into smaller fragments. These can be separately implanted and developed. A third technique of cloning would be to take the nucleus from something like a gut cell and plug it into a fertilized egg that has its normal nucleus removed. This has been done in frogs. These are the three major methods of cloning. All give identical twins.

Moraczewski: It has been pointed out that there would be a time lag. If you took a cell at this present age, you would be that many years ahead of your identical clone, and second, it has been pointed out that the identity is biological and even then not perfectly; that is, even so-called identical twins are not really identical — psychologically, very different. I suppose you could say this was applicable to clones.

Carlson: You could extend this idea all the way to a point of absurdity. Technically and morally, I have reservations about cloning. But, assuming that cloning is possible from a technical point of view, a person could decide to have a clone of himself and that child in turn could then make a clone of himself. Eventually, you could have a whole series, and the person who happens to be in the middle of the series would not only see younger representatives of what he looked like a few years before even to the point of his birth but he would have the dubious pleasure of seeing himself later on in development to senescence and being a crotchety old man. Again these are red herring issues. I don't think these are things people want. I don't think these are things we need fear. I can't conceive of any program where cloning would be a desirable feature in society.

Moraczewski: One has to propose: If one took a genius regarded as giving outstanding contributions to society and cloned him/her, would we multiply the productivity of that individual?

Carlson: We would multiply the potential for those geniuses to exist as clones in society but each individual in a clone could have a different profession or different activities and values depending on where they were raised in the world. We don't clone minds.

Moraczewski: Another question: In counseling, if a counselor miscalculates the outcome of a pregnancy, could the counselee take the counselor to court?

Shaw: Yes. Anybody can sue anybody for anything. It may be thrown out of court.

Walters: Dr. Shaw, wasn't there a successful suit for failure to diagnose phenylketonuria (PKU)?

Shaw: This was a very strange suit and it was successful. The fact situation occurred in Detroit in 1960. Two pediatricians saw a child in the early months of development, 2-3 months of age. The mother was concerned that the child was not developing normally and the pediatricians were very reassuring to the mother that babies have different rates of growth and development and this was a normal variation and not to worry about it. And as it continued, she finally took the baby to another doctor who, on ruling out all the different causes for mental retardation, did a PKU test and found it positive. At this point it was too late to put the baby on a restrictive diet that helps to prevent mental retardation, so the parents sued the original pediatricians. An $800,000 damage suit was awarded on the basis of erroneous diagnosis. It was not on the basis of erroneous genetic risk information.

Moraczewski: Is there any relation between chronic depression and genetics?

Carlson: I wonder about that. I read Julian Huxley's memoirs and he describes that in his family there has been a history of chronic depression. His brother Trevor hanged himself. Huxley himself has had about two or three nervous breakdowns in his time, and I believe his grandfather, Henry Huxley, also suffered from depression. One of his aunts also was manic depressive and was institutionalized. It is a family having an unusual number of talented individuals like Aldous and Julian Huxley; Leonard Huxley; and Anthony Huxley, a Nobel laureate. It is a steller type of family but it also carries along with it, at least in some members of the family, a history of depression. I don't know if this trait has been studied extensively.

Murray: Chronic depression, *per se*, to my knowledge, has not been demonstrated to have a clear-cut genetic etiology. But in manic depressive psychosis, there is very strong evidence that this may be an X-linked disorder, perhaps X-linked dominant disorder. There are a lot of studies that are consistent with that mode of inheritance. But much more needs to be done before we can be certain that this is the genetic mechanism.

Moraczewski: What is meant by a sex-linked genetic trait?

Murray: One should distinguish between sex-linked and X-linked because an X-linked trait, a term I try to adhere to because it's more specific, is a trait determined by a gene carried on the X chromosome. A sex-linked trait can also mean one that is carried on the Y chromosome because it, too, is a sex chromosome. So one should make a distinction between the two in order to be precise, although traditionally when one says a trait is sex-linked, they usually mean the X chromosome.

Carlson: And that should be distinguished from sex-limited inheritance, just to confuse you some more. A male who has pattern baldness has nothing unusual about the sex chromosomes. In fact, pattern baldness is autosomal, which means the gene responsible for the trait is on a chromosome having nothing to do with sex but it happens to be a gene whose effects are very responsive to male hormones. If the male hormone is present, the person with the pattern baldness gene will show baldness, whereas if that same individual has that same gene but has surgical removal of the testes for a tumor as a child, that child, even though potentially pattern bald, will not show pattern baldness. Similarly, women who have some sort of hormonal imbalance in abnormally functioning adrenal glands may release a male hormone and then suddenly find themselves going bald (and yet growing a beard!).

5/
General Comment

Moraczewski: A number of issues have not been touched on or at best touched on very lightly, especially the issue of values. First I want to take up a general question that has been addressed in various ways to me and to all the members of the panel. I will pose it in this way, which is a collapsed form generated from a number of similar questions. It has to do with the question of scientific versus humanistic determination—scientific values versus humanistic values. And the thrust is that some of the comments made in the preceding chapters sounded as if we were placing scientific or technological values prior to humanistic and religious values. As you have already deduced, the members of the panel represent different philosophies and different religious points of view, so there is no uniform answer, obviously, from the members of the panel. The question has to do with the notion of the quality of human life versus the value of human life itself. For example, when it has been determined that a fetus has a genetic disease with varying degrees of malformations, should it be aborted rather than allowing it to survive and experience a diminished quality of life? So the conflict has been put: the quality of life versus the value of life itself.

Carlson: I think both values of life are taken together when parents make decisions or when counselors provide information. Certainly, in my own life, both values play an important part. I can give you one specific example of how my wife and I arrived at a value decision. In 1964 when we had a child, I knew immediately when the obstetrician came in that something was wrong because I had seen him a couple of times before when he signaled success with a little blue button or pink button announcing the news. But from the expression on his face I was quite alarmed. He said, "Your wife's OK, but, sit down, your child has quite a lot of problems," and he started to describe some. He told me that her eyes were quivering back and forth; she had a highly arched palate, a rudimentary jaw, and low-set ears; and she was a blue baby. After he described these symptoms, I knew these were multiple birth defects and was quite concerned. At the time I did not know what was wrong with the child but I suspected there might be a chromosome defect and asked him if there was some way we could check that out. We used the hospital library to try to check out as many references as we could and we weren't able to find the nature of the defect. I recommended that the baby be transferred to the UCLA pediatric ward where I knew there was someone who could do a chromosome count on the baby. In the meantime, my wife and I discussed what we could do. We were prepared for many of the burdens that we could accept for the child. We would have gone through plastic surgery. We would have gone through pediatric cardiology, but when the physician at UCLA brought in the diagnosis of Edwards' syndrome, caused by an extra number 18 chromosome, I realized the hopelessness of the situation. After reading all the available literature we could get and seeing what was ahead for us—a retarded vegetative life with failure to thrive and our daughter's likely death within a year—I said to the physician, "Do me one favor, don't do any heroics." I felt that this was a decision based on our concern of the impact such a child would have on our family. We did not want the child hooked up; we did not want surgery performed as this would cause anxiety and expense for a condition that had no possibility of a cure when so many organ systems were involved. In any of these tragic situations, it is not a callous cost-benefit analysis—it is a time when tears roll down your cheeks; you are aware of your humanness, you are aware of a crisis. You feel under certain bad situations that you will live with it and do everything you can. In other bad situations you just judge it to be hopeless and let nature take its course and let the child die. I think that is a humanistic decision and not a scientific one. I think that is what the two value systems are about. You need the information and you need your emotions to run along with your mind.

Murray: I think that one of the reasons for that question is the fact that people have come to expect so much from science, perhaps much more than science can promise. And scientists, possibly in their zeal to do research and to uncover mysteries, have made promises they have not been able to keep. In my experience in counseling, the majority of people's decisions are not based primarily on scientific information. That information is taken into account but their values and their desires play a role in making a decision. Certainly, we tend to emphasize science and technological application because of our belief that it will help people's decisions. But the tendency is for people's decisions to become harder, not easier, as scientific technology is further and further developed. In the final analysis, rational people — and we call ourselves rational — really act at the gut level when the chips are down. And I think it will be a long time before we stop acting at that level, despite all the new scientific information we get. So even though we emphasize scientific advance, scientific technology, and applications in human affairs, I think all of us recognize the fact that we have a long way to go before we are like Mr. Spock in *Star-Trek* and make our decisions totally based on cold, scientific logic.

Walters: On a theoretical level, science and values cannot possibly conflict because they involve two different enterprises. Science is the development of information by abstracting out certain things and looking at little segments of reality, at smaller and smaller pieces. It's usually an analytical enterprise that seeks to develop factual information and knowlege. It supplies some of the raw material on which we then base our value judgments. I don't see how science and values can conflict if they are in two different universes of discourse. One provides data on the basis of which we make decisions. As we make the decisions, we are inevitably involved in making value choices on the basis of priorities that transcend science.

Moraczewski: I don't know if I would entirely agree with Dr. Walters. Another analysis may be made because science ordinarily, as we understand it, is concerned with *understanding* of how something came to be, how it functions, whether that something is inanimate in nature — the moon, the sun, the galaxies, or whatever other heavenly bodies you might want to consider. On the other hand, technology is concerned with *controlling* and modifying that nature. Of course, this is applicable to the scientific understanding of man and the technological control of man that we have been talking about.

In contrast, values, which are usually encompassed under the notion of *human* values, originate from a different source. They are not necessarily derived in the same way scientific values are. And that's why I think there can be conflicts; that is, the human values originate from a reflection on a different scale, some from a philosophical reflection and some from religious considerations, but they pertain to man as man and not as scientific enterprise. So I think there is a potential conflict of values between the scientific priorities and the values based on a humanistic analysis when these cannot be simultaneously possessed. This leads us to the second half of the question: Who will determine the social and political priorities for the genetic future? Will it be a group of scientists and doctors speaking for the public interest? Or a politically minded few? Or some other group as yet undefined? How would such genetic priorities be voted on at a referendum? Would that be the way the total society would come to grips with the problem? This seems to be one of the difficulties: If those persons of knowledge and goodwill differ about what is important, how is the priority of values to be assigned? What is the quality/quantity type of conflict? Abortion is a good example. It's certainly one of the most heated issues in our nation today. There are persons of equal knowledge, intelligence, and integrity on both sides of the issue. Where do they differ? They differ apparently because of the different priorities of value each group has. What makes it particularly regrettable is that each group thinks it is making the ethically responsible decision, frequently accusing the other of being ethically insensitive. How to resolve that problem is a major task for our society.

Now there is reference to the question of genetics, the kind of individual we want and how much control. Should mass screening for genetic defects be compulsory? Voluntary? A combination?

Murray: This is like so many other questions asked. The answer will depend on what disease you are talking about. The answer cannot be given for genetic diseases in general. The Academy of Science spent two years reviewing the question. The general decision was that there was only one condition — that of phenylketonuria (PKU) — where the community would be ready, where the knowledge of the disease would probably be ready, where knowledge about the therapy was probably ready, and where mass screening might be advocated. Even here, there were problems about delivery of care, about delivery of knowledge, about long-term follow-up that needed to be answered. With that exception of PKU, there are probably no other genetic conditions where all the factors are correct or right, where mass screening would be mandated. This is not to say that voluntary screening should not be available for people educated about the benefits and the harms — the boons and the banes, if you will — that can accrue from entering into the screening program. The screening program is something like being on a giant slide. We are at the very top of the slide, just getting started, and haven't built up too much steam. At the beginning you might be able to stop yourself and climb back up to the top, but once you're half way down, there is no way you are going to get back up. You have to go all the way to the bottom, and the problem with the slide is that you don't know where it ends. It might end on a comfortable pillow by your getting the information that you don't have the trait and your children are not at

GENERAL COMMENT

risk for a disease, or you might be in the situation where you're a member of a high-risk group and at risk of having an affected child or of being affected yourself. You didn't know that when you started at the top of the slide. So for this reason, for the foreseeable future, and I think the National Academy of Science's screening committee would agree, mass screening should not be pushed, and screening programs should definitely be voluntary and not mandatory.

Shaw: I would prefer to see all screening voluntary. But even where the good, as in PKU, so outweighs the bad, the only way voluntary screening would work is by such widespread education that mothers would be clamoring for the test to be performed on their newborn babies. Until we reach that stage, I think offering screening in a mandatory way, even where help can really be given, is not ethical. I think some people are really pushing for mandatory screening. However, the statutes have a conscience clause so that one could say it's against his or her religious principles to be screened. So there is room for individual preferences.

Carlson: I prefer the mandatory screening of infants for conditions like PKU, because the present public awareness of the nature of the disease is so small that without mandatory screening there would be no way of helping these children. I do not favor mandatory screening for adult carriers. I prefer that carriers know of their potential defect from a voluntary program, such as that sponsored by the Hillel groups and others carrying out screening for Tay-Sachs disease.

Shaw: Dr. Murray, why was galactosemia not included in the mandatory group?

Murray: I think primarily because of a cost-benefit ratio with the size of the population that needed to be screened and also because there was some question about heterogeneity in the enzyme deficiencies that cause galactosemia.

Walters: I think a great deal depends on the ethical justification that we would give for mandatory screening. In the case of PKU, mandatory screening is justified on the basis of the best interest of the child who is being screened. It is a far different argument to support or recommend mandatory screening when the benefits accrue primarily to the society as a whole or to the gene pool. I think a social justification is a qualitatively different argument. The only point at which I can see some problem, at least logically, is in the case that

Dr. Murray brought up. What if you had a choice between the extinction of the human species and some kind of coercive screening program? I see that as a very difficult ethical conundrum.

Carlson: I have been given some difficult questions to answer. First, "How do you respond to the criticism that geneticists play God?" All of us in many aspects of our lives are forced to play God. We are forced to make decisions when they are thrown on us. To fail to make decisions is cowardice, an act of hopelessness, an abandonment of responsibility. Was it playing God to have scientists in the nineteenth century advocate public immunization programs to prevent infectious diseases? You could argue that immunization is unnatural because it interferes with the natural order of things. But most people did not look upon natural death from infectious disease as something they would accept. The only time you run into difficulty about playing God is when you forget your humanness, when you don't think through the long-range implications of what you are doing. I feel that this is not playing God but lacking an ethical or moral attitude that should be given to your research. If a scientist, for example, is working in genetic engineering, and he is putting tumor genes into bacterial cells, I believe he is doing a moral wrong if he does not take into consideration accidental contamination of the population. He must consider the possible abuse that will come if his techniques are widely known. I believe scientists need to be reprimanded when they do not think through the implications of their work.

A second question poses two choices in regard to an abnormal child. "One choice is to take the 'scientific' attitude and prevent it from coming into being, and the other is to look at its life and its relation to you as increasing the kind of love, sensitivity, and care that develops in response to its needs, and the question is, Do you have any room in your philosophy for this growth to take place? Psychologists say growth comes in crisis situations. You seem to make no allowance for such possibility of growth." On the contrary, I believe that the decision to abort after amniocentesis or the decision to let a child die when it is in a hopeless condition is an act that requires a feeling of sensitivity and coping with human tragedy. I don't feel that one has to accept an abnormality of catastrophic proportions as something inevitable if one can prevent it. The situation that parents face in making a decision for prevention presents many of the ethical conflicts that they would face if they decide to let the child live. Perhaps the difficulties are not so acute and perhaps not so prolonged in electing an abortion but it is a life decision that they have to live with and that affects their sensitivity to other human beings.

The third question is: "No concern has been expressed about the possibility of the mutation rate increasing due to increasing exposure to mutagenic agents. How can we limit this source of increasing the genetic load?" We should definitely be concerned about agents such as ionizing radiation or chemical additives in our food and environment and take a very aggressive stand to try to minimize the risks in our lives. I think a geneticist's obligation is to make this knowledge public and to participate in arguments about what type of industrial or federal screening programs should be devised to test out the thousands of chemicals that are added to our diet and environment and see which ones cause mutations and which do not. We have to be vigilant in asking our legislators to protect us. We should not sit back and accept all these industrial and commercial novelties in our life and use genetic counseling as a way out. Rather we should prevent as much new mutation as possible from occurring in later generations.

Murray: Dr. Shaw was involved in a publication in *Science* that addressed itself to mutagenicity. I would like to hear her comments relative to that question.

Shaw: A large committee of 17 people, called the "Committee of 17," addressed the question. The recommendations were that the entire field of chemical mutagenesis be studied with the same kind of care and depth that radiation mutagenesis had been studied in the fifties when the Hiroshima and Nagasaki bombings caused great interest among scientists. In fact, it was recommended that some steps be made to try to quantitate the mutagenicity of various chemicals and air and water pollutants in terms of chemical units as we use rads and rems for the physical units of radiation. Some hearings were held in November 1975 in the House, in Congress, dealing with the whole question of environmental pollution. Questions have been asked of the American Society of Human Genetics to find what the input might be for the screening of industrial workers for the possibility of extra chromosome breaks and other damage due to industrial pollutants. One of the concerns was that again we are invading the privacy of the individual if we include him in a mandatory screening program under the employer's requirements, and that we are getting into the bind of stigmatization of individuals who may have abnormal results.

Moraczewski: Dr. Murray made a statement earlier that genetic counseling is like medical counseling on a one-to-one basis, but who is the patient with whom you are dealing, the mother or the fetus? Is the mother the patient and the fetus the problem, especially if the fetus is defective? Perhaps this accounts for the solution frequently proposed: Eliminate the fetus. This is the only situation I know in medical treatment where the disease is treated by eliminating the patient. I wonder whether there is a conflict of interest for the genetic counselor who must deal both with the woman whom he faces and hears and with the hidden individual, the fetus. Who is really the patient for the genetic counselor?

Murray: Dr. Kurt Hirschhorn tried to solve that by saying that the patient is the maternal-fetal unit. Or was it the fetal-maternal unit? I've forgotten which but it was one or the other. In other words, they are tied together in a sort of union; however, since one cannot communicate with the fetus one has to communicate with the mother and try to deal with that person with whom the physician can communicate. At one point I did a little scenario on a physician's or a counselor's attempt to communicate with a fetus in a variety of ways. All that the fetus said was glug, glug, as it swallowed amniotic fluid that was circulating through the sac. I think part of the basic problem is the problem of communication and conscience, the problem of whom you are going to face when you look into the eyes of the patient—into the eyes of the mother in this case. One does not face the fetus, it's the mother, and in the classical, the traditional medical model, you must care for the person who sits in front of you, before you, or beside you, and not for the person that is not seen. Otherwise, the whole basis for our current concept of doctor-patient relationship must be drastically revised.

Shaw: I would give a human interest experience on this problem of looking into the eyes of the mother and talking with her. Dr. Jerry Mahoney of Yale University told about the time he had first used a fetoscope to look inside the uterus and see the fetus. Just as he found the hand, the hand found his instrument and battered it down. This gave him an emotional feeling that he didn't know possible in terms of human response to the fetus.

Walters: I was going to ask, in terms of prediction, whether you think the direct visualization of the fetus might have the effect of making us think of the fetus more as a child than we currently do. Currently, it is invisible to us.

Shaw: I don't think that you could help but go in that direction, and if you had artificial wombs and could look at the fetus the whole time, it would be even more so.

GENERAL COMMENT

Murray: The analogy is being made between the ability to visualize the fetus and the ability of the bomber pilot in warfare, 36,000 feet up, to see the people on whom he is dropping bombs. He doesn't see the agony and the horror on their faces as they are destroyed, as opposed to the hand-to-hand combat on the battlefield and the effect this has on the man or woman who kills "the enemy" and sees the effect of what they do. There again are gut feelings not related to the scientific facts or knowledge of physiology or chemistry.

Carlson: Dr. Murray has stated that, if he had the choice between some sort of scientifically controlled genetic program where the reproductive capacity of every human was scrutinized and selectively permitted to go through or aborted and seeing a more humane and loving world in which the genetic load kept increasing to the point of extinction, he would elect for extinction with values rather than the preservation of life with compulsory attempts that deny these human values. It reminded me of a House Un-American Activities Committee at which H. J. Muller was asked some questions about his experiences in the Soviet Union in the 1930s and he made the remark that if he had a choice of seeing the world live under the conditions that he experienced in the Soviet Union during the Stalin terror or seeing the world annihilated by atomic war, he would prefer seeing the world annihilated by atomic war. I disagreed with Muller for all the respect I had for him as a teacher because I felt that even if it were possible for the Soviet Union to take over the entire world, dictators come and go and totalitarianisms come and go, and there is a quality that keeps popping out from human beings, no matter how adverse their situation is, to effect changes, to restore values, so that eventually you're free of totalitarian regimes. And I was wondering if a similar situation doesn't exist parallel to what Dr. Murray was drawing. Given an abnormally high genetic load, and given the compulsion to interfere, wouldn't it be possible in the long run and wouldn't it be better in the long run for humanity to go through a period of crisis in which human rights have been infringed while humanity prevails with the hope that it can then restore its values rather than to see humanity lost? Why do you prefer the loss of humanity?

Murray: Partly because I don't see humanity as we know it as necessarily the highest order of creation. Perhaps we are merely a stepping-stone to another organism, or creature, or being that has more to contribute to the world, the universe, or what have you. Also because I am sure that there are civilizations that may have been lost because of the kind of regimentation and control that I fear. I have an inherent belief, if you will, in the mechanisms that have led us to evolve to the state where we are. These operate to balance the damage we are supposed to be doing to ourselves. So, finally, it is because I do not fear that the loss of this civilization—this species on this planet—would necessarily bring an end to mankind because of the probability of many, many thousands of other living experiments going on in other galaxies across the universe.

Carlson: Small comfort for us certainly.

Murray: Of course, I'm here only for a very short time and perhaps it is selfish for me to say I won't be here any more so I don't care.

Moraczewski: I hope my antennas are not showing?

Shaw: We received a very interesting comment from the audience and I am so pleased someone could help me out. Quote: "Regarding the Roman medieval law, customs and fetal rights, consult John Noonan's bibliography in *Contraception* and Raymondson's *Women in Greek and Roman History.* In general a pregnant woman received special handling, suspended sentences, etc. The fetus was not the subject of legislation." Written by a history teacher from the University of Michigan.

Moraczewski: One item in regard to Dr. Murray's response made me think of martyrs in various cultures who were willing to die rather than yield to certain values. And perhaps that might be the same. If the human race were suddenly face to face with a certain situation, which to continue in existence would mean to accept values that were totally inimical to it, then a kind of racial martyrdom might be called for. I don't think it will take place but it is a way of emphasizing that there are values, however, which many accept, that do transcend the mere value of life itself, individual or social.

Carlson: A friend of mine was in a German concentration camp when he was around 15-18 years old. As conditions worsened, he spoke to a rabbi and said that he was tired of eating rats and living on garbage and saw no point to life. Why should he continue? The rabbi (who later died in the camp) told him that many, maybe all of them, would die, but conscious life itself is beautiful and one has to go on. This will pass. Humanity is worth preserving and that attitude should prevail over the idea that certain values are more important than conscious life itself. My feeling on what the ultimate tolerance should be in increasing the genetic load is to intervene rather than to allow extinction.

Moraczewski: If life prevails, then why are we so quick sometimes to abort? It becomes a case of selective application of a common value, which can be twisted around to meet particular needs. That's a danger we face.

Another question was asked — a very specific one: What types of cancer are genetic, if any, and what kinds are environmental? In other words, is there a relationship between genetic components and cancer?

Shaw: I'll refer you to my friend and colleague in Houston, Dr. Alfred G. Knudson, who has written extensively on the relationships of genetics and cancer. There are some cancers that follow Mendelian inheritance and I'll just name a few. One is multiple polyposis of the colon. There are thousands of polyps in the colon. These polyps undergo malignant deterioration or change, so cancer of the colon ensues usually before the age of 40 and the patients die of cancer. A variant of multiple polyposis is called Gardner's syndrome where there are polyps in the colon and also some pigmentation around the lips and there are little tumors under the skin called fibromas that are benign but can also develop into fibrosarcomas, which are malignant. So if you had this gene you might die of a sarcoma or carcinoma. These are two different kinds of cancer that are genetic. One that I am sure you have heard about, because it occasionally gets attention in the newspaper, is retinoblastoma. This is a cancer of the retina of the eye and the parents are faced with the choice of having one or both of the child's eyes removed (since it often occurs bilaterally) and having the child blind for life, or letting the child die of cancer. They are sometimes able to remove one eye and irradiate the other eye. But the radiation may lead to blindness. These are very simple one-to-one gene relations. However, the theory is being proposed by Dr. Knudson that cancer is a series of steps and changes in the cell and the genetic change that we talk about and that is inherited is the first step out of many. If this first-step change is in the germ cell, it appears like a Mendelian genetic trait; if it appears in a somatic cell, a breast cell, or prostate cell, it is the first step to that cell becoming a cancer, but it will not be passed on to the next generation. There have been some calculations that some types of cancer are the results of only two changes in the cell; others, three changes in the cell. This does not eliminate the theory of viruses causing cancer. The virus may be one of the steps. There is a lot of thinking going on about genetics and cancer.

Moraczewski: The question was asked about current experiments dealing with bacteria and viruses involving DNA recombination. Do the members of the panel support or are they against a moratorium on such experimentation with DNA?

Carlson: If experiments are done by splicing genes from one organism into an entirely different organism, I would like to see the stringent precautions that existed in the later years of the nuclear reactor development where this was handled with many safeguards against accidents taking place. One suggestion raised on this issue was to use something like the early lunar landing contamination procedure and have individuals who want to do this research do so in specially equipped laboratories, applying for special grants to travel to these facilities and to have the special supporting personnel that minimize the risk of accidental contamination. Why is there this fear? Suppose you investigated some of the genes that are associated with cancer and identified them and spliced them into a bacterium that has been known as a standard for all genetic research for the last 30-40 years—the colon bacterium. If this tumor gene is spliced into the colon bacteria with the good intent of studying the product of the tumor gene and getting out some sort of answers to the cancer problem, what will happen if the person is careless and these bacteria go down the sink into the sewage, get carried away and are multiplied, picked up in the drinking water, and begin to pass from gut to gut so that several million people harbor these bacteria with the tumor gene in it? You won't know about it for 25-30 years because it takes a long time from an initial introduction of one tumor cell to the time it becomes a full-blown detectible cancer. That way you wouldn't know that there is an epidemic that you created until almost a generation has elapsed and by that time it is so overwhelming that you won't know what to do about it. That's the worry, a legitimate worry. I personally don't think this is something toward which one should take a light attitude. Unfortunately, many of my colleagues, self-policing or not self-policing, consider the whole thing a fabrication not to worry about, and they will go right ahead and do this work. I would like to see some sort of restraint put on it even to the point of removing grant funds for doing this research so they would have to fund it out of their own pockets.

Murray: I agree with Dr. Carlson's concerns. I think they are real and

GENERAL COMMENT

I think an additional point he has made is that these are expensive — extremely expensive — experiments, and also that the facilities that would have to be devised to protect us from getting the recombined DNA out in the population are also very expensive. We might consider using the same money in the kinds of research on environment, pollution, and consequent genetic damage and also on research on environmental influences on gene expression in a much more beneficial fashion. At least for the immediate future. One of the counter arguments presented, however, is that the experiments be done on bacteria and viruses that grow on frogs, some species far removed from human beings and also using genes that are not malignant but genes for other factors not causing any infectious disease.

Walters: I wish to raise a question about the parallel between genetics and other fields. I'm wondering if Dr. Carlson would think that the dangers of genetic research are a more serious problem than, say, a lethal toxin that might escape and might kill hundreds or thousands of people. Is this a qualitatively different problem because it is genetic?

Carlson: Yes, it is qualitatively more serious because when you have an organism that can multiply and spread you have no way of knowing how to contain it. Something like a toxin that is released like a gas or other noxious substance doesn't have the biological property of replicating itself. One bacterium in a room may be enough to do the damage. That's the worry.

Moraczewski: One query refers to a statement made earlier by Dr. Murray relative to Down's syndrome and this involves the long-range psychological effects within the family where one child, the first child let us say, has Down's syndrome and is still living. Two questions now: What would be the effect on the first child if the parents signify a willingness to abort a second child who might have Down's syndrome, and what is the effect on the child who did not have Down's syndrome and found out that the parents had considered to abort it if it had the extra chromosome?

Murray: Dr. John Fletcher recounts stories in which this has actually taken place, where there is a child in the family with a certain condition. I don't recall if the child had Down's syndrome or not. However, the child had enough intellectual function so it could understand that mommy was going to the hospital to get something and that she came back without it. The mother was pregnant; she went to the hospital and, when she came back from the hospital without the baby (because apparently the child was defective and was aborted), the child was hiding in the closet and refused to come out and greet the mother and would have nothing to do with her. Apparently, the child understood that the mother had gotten rid of the baby and feared, not seeing the kind of parallels we can draw between the fetus and living individual, that the mother was now going to get rid of the child. This is the kind of effect that might result. Whether this would happen in all children, whether the mother should have talked to the child ahead of time to explain and tell him what was going on so that the child would not misunderstand and think she was going to get rid of him, I don't know. It can have this kind of effect, even when a child, as apparently in another case, was not aborted asked the question, "What if I had been defective, would you have had me aborted?" And the answer to the question, of course, is yes. But this mother answered "I don't know." These are hard questions and if one is face to face with such questions, it's very hard emotionally to deal with them. Under pressures of the anxiety and the hostility that one can feel when one is carrying such a child and has the opportunity not to have it, actions can be quite different. In any event I don't know of any studies more than anecdotal reports like these I have just mentioned where this has been looked at in any kind of systematic fashion to determine whether indeed this does have an effect on the child already in the family.

Shaw: I am reminded of the first large family of familial mongoloid studies in 1962 where there had been six mongol children born. The woman whom I was counseling had one mongol who was 5 years old and she was keeping it at home. Now she was pregnant again. This was before the days of amniocentesis and she understood her risk of about 20 percent that the child would be a mongoloid because of a chromosomal translocation. She said to me she had received a lot of understanding, comfort, security, and empathy from the first mongoloid child she had and that she looked on her 5-year-old boy like she looked on her normal 8-year-old daughter and her 9-year-old daughter. All of them would be in the life of the family for a temporary length of time, then they would fly the nest. With her older children it would be when they went to work, when they went to college, or when they were married, and with the mongol child it would be when he was 6 or 7 years old and she could no longer care for him at home. This was the way she was looking at her present pregnancy. So I think there is a great diversity of feeling and attitudes toward these things. Perhaps some of it is from some of the tragedies we have already experienced in life that help us to prepare. With some people, their experience makes them less able to cope; and with some, their experience makes them better able to cope. I think there is a tremendous variability in how people look at this problem. I know a man with a mild deformity who was asked a question about whether he would advocate amniocentesis and abortion. The question was: What if your mother had aborted you for having club feet, how would you feel? He said, "I wouldn't know about it and so it wouldn't matter to me at all—it wouldn't make any difference." So that's one way to look at it. Another person might be devastated by the thought that his mother might have aborted him. I think we have such a diversity of opinion and diversity of values.

Walters: Insofar as abortion is a serious moral issue, I'm wondering if we have really emphasized enough the possibility, even the obligation, to prevent avoidable defects preconception. When it comes to a choice between counseling a couple not to have children, perhaps disappointing some of their aspirations on the one hand and likely abortion on the other, it seems to me that the balance should be tipped, generally speaking, in favor of not conceiving. And perhaps we haven't talked strongly enough about what Paul Ramsey calls the ethic of genetic duty. If one is 40 years old, and the likelihood of bearing a child with Down's syndrome has increased dramatically, perhaps the ethical duty at that point is not to expose a potential child to the risk of being born with Down's syndrome. Perhaps when carriers of a recessive trait marry, we ought to emphasize more strongly than we have earlier the possibility of this couple's not bearing children rather than risking the production of a seriously deformed child or the possibility of abortion for genetic reasons.

Moraczewski: This last question refers to Dr. Walters's presentation in which he cited three cases: the first case was artificial insemination; the second, *in vitro* fertilization; and the third, *in vitro* gestation, in sequence. The question pertains to all three: What is the teaching of the Catholic Church in regard to these three cases? *In vitro* gestation has not been treated by the Magisterium as far as I know. The question of *in vitro* fertilization was briefly addressed by Pope Pius XII who concluded it was "immoral and absolutely illicit." Artificial insemination had been treated as early as 1894 when the question was put to the Sacred Congregation of the Holy Office. The question: May artificial insemination be used on a woman? The answer: no. No explanation was given, which doesn't help an ethicist who is trying to understand. Pope Pius XII, who wrote most extensively among the popes in regard to medical ethics, did consider that issue and emphasized again the prohibition in regard to artificial insemination. As he put it, it was a defect in the conjugal relationship between a husband and a wife. The natural law is such that human offspring are to be formed by means of marital intercourse and only by this means. Many would hold that teaching to be reformable; that is, it can be changed. Whether or not it will be depends on new facts — if such there be — and on further reflection on the values to be promoted.

I hope we have uncovered many of the problems, given different aspects and different sides, even though they may have been incompletely presented. We have tried to give an insight into some of the stirrings in the area of genetics, genetic counseling, law, and social concerns. Many problems face us, especially in a pluralistic society that does not have an agreed-upon value system. It is a challenge to all, no matter where we stand on the issues, for us to come to some kind of agreement that would be best for those concerned.

Epilogue

IN THE PRECEDING chapters we have indicated something of the magnitude of the genetic problems facing the Tricentennial People. Select individuals from that category—the students of today—have indicated with incisive comments and probing questions some of their concerns. It is too soon to sort out the key questions, the basic issues, the heavy anxieties. Yet, I will hazard to select a few items for emphasis here—items that appear to me as meriting special attention in the coming decade.

There is a need to assess accurately the *multidimensional* impact of genetic disease on our society. Merely to identify the medical consequences is not sufficient basis for the formulation of public policy. It also would be necessary to determine the impact of genetic disease and its treatment, or nontreatment, on health care delivery systems, on basic rights and values of the concerned individuals, on religious beliefs and ethical convictions, on interracial relations, on the economic health of families and society, on the self-image of those who are physically and mentally below "specifications," as well as on the self-image of their parents. Furthermore, it would seem that a consensus first must be reached as to what degree of genetic disease would be acceptable to society. It is not at all evident that the biology and sociology of genetic disease are yet that well articulated or understood.

ALBERT S. MORACZEWSKI, O.P., Ph.D., is President of the Pope John XXIII Medical-Moral Research and Education Center, St. Louis, Mo. He is an Adjunct Associate Professor of Ethics, St. Louis University, St. Louis. He has been Associate Professor of Psychopharmacology and Coordinator of the Career-Teacher Training center in Addiction at the Baylor College of Medicine and also Professor of Theology and Science at the Institute of Religion and Human Development in the Texas Medical Center. He was a research specialist in neuropharmacology at the Texas Research Institute for Mental Sciences in Houston. He has published articles in the *Bulletin of the Atomic Scientists, Archives Internationales de Pharmacodynamie et de Therapie*, the *Journal of Neurochemistry, The Thomist,* the *Journal of Histochemistry and Cytochemistry, Thought, Psychosomatics,* and he has collaborated in several compilations of essays concerned with bioethics.

Many people find abortion unacceptable as a "treatment" of human fetuses who have been identified as genetically defective. Elimination of the patient is hardly good therapy! An increased level of support for research is needed in this area to uncover alternative ways of reducing the impact of genetic disease on the afflicted individual, on the family, and on society at large.

Mass screening mandated by government has been proposed as a means of reducing the number of persons with inherited disorders. Further, it is argued that just as we now require the quarantine of individuals with specified communicable diseases, so, too, we should reproductively isolate those who would pass on defective genes. Granted the numerous similarities, there are also some differences: The person quarantined for infectious disease is usually ill himself whereas the heterozygous carrier of a genetic disorder generally is not; the former ordinarily is quarantined for a relatively short time whereas the latter is in reproductive isolation for a lifetime.

There is a fear, not fully articulated, that genetic counseling and genetic engineering may be eventually coupled with a type of social engineering that would legislate coercive compliance with genetic laws. Is it inconceivable in our current intellectual and value climate that the state would attempt to enact legislation directed to determine not only the *quantity* but also the *quality* of children to be generated and/or born into our society? Would, then, public policy override fundamental individual rights in order to produce genetically specified humans? It would be necessary for the state to establish a compelling reason for such drastic action.

There are prior questions, it would seem, that are concerned with the criteria for the determination of socially acceptable or unacceptable behavior. Are we certain that today's values regarding such behavior will be applicable to tomorrow's conditions? The influx of immigrants from other nations to the United States in the past 100 years, and the large exodus of peoples from the farms to the factories in recent decades, have amply demonstrated that values pertaining to human well-being in a rural environment are frequently inapplicable in an urban milieu, and vice versa. Biological traits suitable for optimum performance in one set of circumstances may be totally unsuitable in another. Hence, since the nature of our future social structures is unclear, can we logically and safely legislate *now* what kind of genetically determined human characteristics would be desirable for tomorrow?

Underlying the desire for a genetically healthy population may be a value requiring humans to be born perfect as if they were the output of mass production techniques requiring that each individual item that does not meet certain tolerance specifications be discarded. This could be interpreted as an element in the search for an earthly paradise—a habitat without defect or flaw. B. F. Skinner's *Walden II* provides an inkling of what such a contemporary utopia might be like. Casual readers may detect, at least, a metallic "aftertaste" after the reading of this attempt to construct a modern paradise based on behaviorist psychology.

A significant theological issue arises when we consider the many complex questions related to genetic engineering, particularly under the aspect of redesigning man. Just what degree of autonomy do we have over our own biological destiny? If we grant that the human body, at least, is the product of an evolutionary process, we can ask whether further *biological* evolution is possible, or even desirable. Many would say that we are now in a phase of moral and social evolution, and that such development is under man's control—if only relatively. But is it part of the divine order of things that man assume control over his own future biological development? It seems that here we have a significant theological question, the answer to which would determine a Christian stance with regard to many aspects of genetic engineering.

Yet as awesome as some of these considerations may be, there is no need to look upon these biotechnological advances as necessarily opposed to man's well-being. Attitudinal guidance is provided by two contemporary ecclesiastical statements, both of which strike an essentially optimistic note:

> ... man should, by the use of his skills and science of every kind, acquire an intimate knowledge of the forces of nature and control them ever more extensively. Moreover, the advances hitherto made in science and technology give almost limitless promises for the future in this matter. (Pope John XXIII, *Mater et Magistra*, #189)

> Thus, far from thinking that works produced from man's own talent and energy are in opposition to God's power, and that the rational creature exists as a kind of rival to the Creator, Christians are convinced that the triumphs of the human race are a sign of God's greatness, the flowering of his own mysterious design. (Vatican II, *The Church in the Modern World*, #34)

These reflections are not intended to summarize the rich contents of the foregoing chapters. Rather they are meant to direct attention to a few aspects that, for this writer, appeared to have more than a passing interest. Our genetic future depends in great measure on the wisdom we exercise in the present. Wide-ranging discussions, as the symposium on the human applications of the new genetics, will help to provide a broad base of informed persons from which public policy can be shaped more effectively and justly.

Albert Moraczewski

Index

Abortion, medical, 10-11, 14, 37, 53-54, 57, 71, 84, 96
Abortion, spontaneous, 4, 23, 49
Adoption, 19-20, 25
Albinism, 6
Amniocentesis, 10-12, 14, 23, 37-39, 44, 52-53, 65, 69, 71
Anencephaly, 12
Artificial insemination by donor (AID), 12, 20, 25, 68, 74, 95-96

Birth defect, 4

Cancer and genetics, 91
Chimeras (humanoids), 15
Cloning, 14-15, 77-78
Confidentiality, 52
Cooley's anemia, 43
Cri-du-chat syndrome, 49
Cystic fibrosis, 6, 32, 43, 48

Dight, Charles, 34
Down's syndrome, 7-8, 22, 31, 37-38, 49, 65, 71, 93-95
Down's, translocation, 23

Edward's syndrome, 8, 82
Eugenics, negative, 11, 26, 33
Eugenics, positive, 11, 13, 20, 26-27

Fetus, legal rights, 53
Fletcher, Joseph, 67, 70, 71

Genetic counseling, 8, 13-14, 29-31, 49-50, 77, 87
Genetic engineering, 12

Genetic load, 4-5, 11-13, 15, 18, 21-22, 28, 87
Genetic risk, 47
Genetic screening, postnatal, 70, 72
Genetics, medical, 11, 15, 46
Goertzel, V. and M., 17

Hemoglobin AA, 42
Hemoglobin AS, 42
Hemophilia A, 6, 40, 49
Heterozygotes, 7, 21
Hirschhorn, Kurt, 88
Homozygotes, 21-22
Huntington's chorea, 6, 13, 24-25, 32, 49, 62-63, 75
Huxley family, 79

Infectious vs. genetic disease, 55-56, 61-62
Inovulation, artificial, 20
IQ, 17, 26

Kass, Leon, 67, 74
Kleinfelter syndrome (XXY male), 11, 39
Knudson, Alfred G., 91

Lesch-Nyhan syndrome, 39, 57
Life expectancy, 6

Mahoney, Jerry, 88
Mandatory screening, 42, 44, 60
March of Dimes, 76-77
Medical advising, 29, 33
Miscarriages, 24
Mongoloid idiocy. *See* Down's syndrome
Muller, H. J., 27, 89

Muscular dystrophy, 49
Muscular dystrophy, Duchenne's, 40
Mutagenesis, chemical, 87
Mutations, 4–5, 14, 58, 74

Normalcy, 19, 27

Orchomenos, tragedy of, 42
Overdominance theory, 21

Patau's syndrome, 8
Pattern baldness, 25, 80
Person, legal, 60–61
Person, natural, 60–61
Person, ontological, 60–61
Phenylketonuria (PKU), 9, 48, 60, 69, 71, 79, 84–85
Prenatal diagnosis, 24. *See also* Amniocentesis

Ramsey, Paul, 57–58, 67, 74, 95

Recombinant DNA, 92–93

Schockley, William, 58–59
Sex chromosomes, 8
Sickle-cell anemia, 6, 9, 21, 32, 34, 41–42, 49, 55, 65, 70
Sickle-cell trait, 41–42
Sperm banks, 13, 20, 26, 68
Spina bifida, 10

Tay-Sachs syndrome, 6, 8–11, 32, 34, 36, 49, 55, 70, 85
Test-tube fertilization, 67–69, 72, 95
Thalassemia major, 43
Tort law, 53
Trisomy 21. *See* Down's syndrome
Turner's syndrome (XO), 11, 39, 49

X-linked disorders, 39–40, 80
XYY anomaly, 11, 39

THIS VOLUME is an outgrowth of a symposium sponsored by Clarke College, Dubuque, Iowa, October 21, 1975. The symposium was centered on the biological, legal, moral, and psychological effects of the use of the new genetic information. It was one in a series of Clarke-sponsored symposia on contemporary topics including "Liturgy and Architecture" in 1964, "Creative America" in 1965, "Man in a Man-Made World" (effect of technology and behavioral mechanisms on the human person) in 1968, and "Creative Dissent" (political currents of the 1960s) in 1971. Clarke offers these symposia as a service to the college communities of Dubuque and to the city. They bring together scholars who engage in open interchange on topics of immediate concern and awaken students to the complexities of the larger world.

ISBN 0-8138-1650-5